Spirit and Nature

An Interfaith Dialogue

SPIRIT and NATURE

Why the Environment Is a Religious Issue

Edited by Steven C. Rockefeller
and John C. Elder

BEACON PRESS / BOSTON

Beacon Press
25 Beacon Street
Boston, Massachusetts 02108-2892

Beacon Press books
are published under the auspices of
the Unitarian Universalist Association of Congregations.

Printed in the United States of America

99 8 7 6

Text design by David Bullen

Library of Congress Cataloging-in-Publication Data

Spirit and nature : why the environment is a religious issue : an interfaith dialogue / edited by Steven C. Rockefeller and John C. Elder.

p. cm.

Essays originally presented at a Symposium on Spirit and Nature held at Middlebury College in 1990.

Includes bibliographical references and index.

ISBN 0-8070-7708-9. — ISBN 0-8070-7709-7 (pbk.)

1. Human ecology—Religious aspects—Congresses. I. Rockefeller, Steven C., 1936– . II. Elder, John C., 1947– . III. Middlebury College. IV. Symposium on Spirit and Nature (1990 : Middlebury College).

GR80.S65 1992
291.1'78362—dc20 91-37116

One thing is all,
All things are one—
Know this and
All's whole and complete.

—Seng-t'san, "Affirming Faith in Mind"

The "brotherhood of man" needs to be widened to embrace not only women but also the whole community of life.

—Rosemary Radford Ruether,
Sexism and God-Talk

CONTENTS

ILLUSTRATIONS

FIGURE

ACKNOWLEDGMENTS

THE essays in this volume were originally presented as part of a symposium on "Spirit and Nature" held at Middlebury College in 1990. In organizing this symposium and in preparing this book, we were assisted by many individuals and institutions. We wish especially to thank our colleagues at Middlebury College, Professors Barbara L. Bellows and Stephen C. Trombulak, who worked with us on a Symposium Planning Committee; Professor J. Ronald Engel, who helped relate the symposium to the World Conservation Strategy project of the International Union for Conservation of Nature and Natural Resources; and Claire Wilson who has been extraordinarily helpful with manuscript preparations, and who together with Janet Winkler provided invaluable assistance with the planning and direction of the symposium.

Many people in the Middlebury College community gave generously of their time and energy in helping with the symposium. We appreciate the support given by Joanne Lopez Hill, Frank Kelley, Ron Nief, Chaplain John Walsh, Lis Grinspoon, '91, and Professors Paul Nelson and O. Larry Yarbrough, as well as by President Olin C. Robison and President Timothy Light. Thoughtful assistance was provided during the planning process by Nancy Nash, founder of the Buddhist Perception of Nature project in Thailand. Richard H. Saunders, Director of the Christian A. Johnson Gallery, Emmie Donadio, Christine Taylor, and Ken Pohlman assisted us with the art exhibition "Spirit and Nature: Visions of Interdependence," from which have come the illustrations for this volume. For their financial support we thank the Christian A. Johnson Endeavor Foundation Endowment, the Nathan Cummings Foundation, Mary R. Callard, Dr. and Mrs.

Richard M. Chasin, Mr. and Mrs. Willard Jackson, Mr. and Mrs. George D. O'Neill, Mrs. Leo Quinn, and Laurance S. Rockefeller.

We deeply appreciate the special efforts made by Bill Moyers, Judith Moyers, and Gail Pellett of Public Affairs Television in preparing an outstanding documentary film that significantly extends the outreach of the symposium. Our thanks are extended to the United Nations Environment Programme for permission to include the United Nations *World Charter for Nature* in this volume. Claire Wilson provided invaluable assistance with word processing, correspondence, and research, and Lauren Bryant and Carol Leslie of Beacon Press have been most helpful throughout the editing and publication process for this book. We wish to acknowledge our gratitude to them.

Finally, we warmly thank all of the symposium speakers for their willingness to join us at Middlebury for four days. Their essays, collected here, contribute strong voices to an urgent and far-reaching dialogue.

INTRODUCTION

THE global environmental crisis, which threatens not only the future of human civilization but all life on earth, is fundamentally a moral and religious problem. It calls upon us to exercise our human freedom with a renewed sense of humility and responsibility. In order to explore constructive responses to this challenge, a special four-day symposium entitled "Spirit and Nature: Religion, Ethics, and Environmental Crisis" was held at Middlebury College in the fall of 1990. Its purpose was to foster ways of imagining and living in the natural world that promote sustainable development, joining scientific understanding with life-affirming moral values and world-affirming religious values. This volume contains the addresses delivered during the symposium, along with certain other selections from the proceedings.

The symposium speakers represented the Buddhist, Christian, Islamic, Jewish, Native American, and liberal democratic traditions. There was also a presentation of the second World Conservation Strategy being prepared by the International Union for Conservation of Nature and Natural Resources (IUCN) in collaboration with the United Nations Environment Programme (UNEP) and the World Wide Fund for Nature. The addresses, and the discussions growing out of them, raised questions that are becoming ever more urgent at this moment of unprecedented danger to the global environment. In what sense does the environmental crisis reflect a crisis of moral values and religious faith? What spiritual resources do the various religious and ethical traditions of the world hold for us at such a time? What do the different traditions have to say to one another today that may clarify what it means to have a proper respect for the earth in our personal and social choices? And how do religious traditions need to be reevaluated and reconstructed in light of our increasing environmental difficulties? The organizers of the Middlebury symposium hoped that it would also help to stimulate dialogue about these subjects in a wider circle. One important contribution to such a broader impact is the public television documentary about the symposium produced by Bill Moyers and first aired in June of 1991.[1] The editors of this volume hope that it, too, will be a helpful point of reference within the growing community of religious and environmental concern.

Human history's major theme in recent centuries has been a struggle to overcome oppression and a quest for social and individual freedom. Despite the persistence of poverty in much of the world and the emergence of new forms of social oppression, gains in intellectual, political, economic, moral, and religious freedom have been substantial throughout much of the world. Nevertheless, this new freedom has not brought the happiness that many believed would automatically follow. There is an important lesson to be learned in this regard, which is directly related to the current environmental crisis.

Freedom in the sense of liberation from oppression and achievement of equal opportunity is a prerequisite for full human self-realization, but freedom by itself does not assure well-being and fulfillment. Rather it confronts humanity with a supreme spiritual challenge. In a free world everything depends upon the capacity of the individual to perfect his or her freedom through faith and responsible action. Freedom without guidance by wisdom is dangerous. It easily becomes destructive to the individual and to the community, or communities, upon which every individual is dependent.

The dangers that come with increased freedom are becoming especially apparent today in humanity's relations with the larger community of life on earth that makes up the biosphere. Thinking that it was headed toward unlimited progress, contemporary industrial-technological civilization is now discovering that it is instead well on the way to catastrophe. Humanity may destroy the possibilities for life on earth unless the freedom and power that we have acquired are channeled in new creative directions by a spiritual awareness and moral commitment that transcend nationalism, racism, sexism, religious sectarianism, anthropocentrism, and the dualism between human culture and nature. This is the great issue for the 1990s and the twenty-first century.

Even though the spiritual life is not confined to the institutional religions, the latter are in a position to play a vitally important role in the process of personal transformation and social reconstruction that is required to address the environmental crisis. If a religion is in harmony with the creative spiritual energies of the times, its myths, symbols, and rituals have the power to touch the heart and awaken faith. Ideas of God and teachings about the relation between God and the world shape human attitudes toward nature. A theology can obstruct development of a respect for nature or foster it. In brief, the cooperation of the world's religions in helping civilization address the environmental crisis is essential. Also religions may recover some of their lost vitality and relevance by reconstructing their own traditions in response to our day's global environmental problems.

The dialogue among religious leaders at the "Spirit and Nature" symposium built upon several decades of steadily increasing concern about our global environmental crisis. The American public has awakened to this crisis through a series of dramatic events, from the publication of Rachel Carson's *Silent Spring* in 1962 and the first Earth Day in 1970 to Bill McKibben's recent analysis of global warming in *The End of Nature,* (1989), the worldwide celebration of Earth Day's twentieth anniversary, and the ten-part public television series entitled "Race to Save the Planet," in 1990. A wide array of private organizations have undertaken scientific research, grass roots organizing, legal action, and lobbying. Groups such as the National Audubon Society, Sierra Club, Wilderness Society, World Wildlife Fund, Natural Resources Defense Council, and the Conservation Law Foundation, and more recently, Greenpeace, American Rivers, and Earth First! have mobilized hundreds of thousands of Americans in efforts to save the earth. In the 1970s, politicians responded with creation of the Environmental Protection Agency and a series of new legislative initiatives including the Marine Mammal Protection Act (1972) and the Endangered Species Act (1973), which established certain legal rights for nonhuman species.

During this same period, writers and religious leaders also began to explore the implications of our environmental crisis and the nature of our present choices. For a decade before the appearance of Carson's *Silent Spring*, a small group of Christian theologians with roots in the process philosophy and theology of Alfred North Whitehead and Teilhard de Chardin had been exploring a new theological understanding of the relation of God, humanity, and nature. Their concerns included greater respect for nature and a new sense of the interdependence of humanity and nature. Publication of *Silent Spring* intensified and gave a new focus to their efforts, and in 1964 the National Council of Churches created a Faith-Man-Nature Group that met annually for ten years. It included among others Philip N. Joranson, Richard Baer, Jr., H. Paul Santmire, and Daniel Day Williams.

The publication in 1967 of Lynn White, Jr.'s, essay on "The Historical Roots of Our Ecological Crisis" introduced a new dimension into these studies of Christianity and ecology. White, who was a medieval historian with deep environmental concerns, severely criticized the Judeo-Christian tradition for generating a view of the universe that sharply separated God and the world and humanity and nature, encouraging attitudes of conquest and exploitation in relation to nature. Other voices joined White's, and Jewish and Christian thinkers concerned about the role of religion in the environmental crisis found themselves on the defensive.

The attack forced theologians to look more deeply into their historical traditions. They could not deny the force of some of the criticism or the need for new theological thinking. However, their further explorations enabled them to identify in both Scripture and certain later traditions the foundations upon which to construct ecologically sound visions of the religious and ethical life. Led by theologians like John B. Cobb, Jr., and H. Paul Santmire, a flood of essays and books in the new fields of environmental ethics and eco-theology followed. One of the most significant early attempts to argue that the basis for a constructive moral and religious response to the environmental crisis could be found in historical religious traditions came from an Islamic scholar, Seyyed Hossein Nasr. In *Man and Nature*, which was published in 1968, he urged a return to a sacramental view of the universe such as one finds in classical Islamic thought and in other traditions as well.

In the 1970s, the professional philosophers too became engaged and by the end of the decade the journal of *Environmental Ethics* had been established. The animal rights movement found new energetic intellectual leadership in the work of Tom Regan and Peter Singer. The deep ecology movement assumed distinct form in response to the work of the Norwegian philosopher Arne Naess in the early seventies. It quickly found strong American support among thinkers like George Sessions and Bill Devall, who had a deep appreciation of the work of Henry David Thoreau, John Muir, Aldo Leopold, and Gary Snyder. The deep ecologists emphasized biocentric environmental ethics and ecological egalitarianism, arguing that the human species should dramatically reduce its population and curtail activities that threaten other species and ecosystems. Many deep ecologists embrace a practice of bioregionalism, which calls upon people to settle in one place and to cultivate a life-style in close harmony with their region.

The feminist movement has developed its own approach to the environmental crisis. Ecofeminism emphasizes the interconnection in western society of the oppression of women and the abuse of nature. Ecofeminists are especially concerned with the critique of power relationships, and they argue that an enduring solution to problems of oppression and exploitation requires a major social transformation that ends patterns of domination and control in all human relations, including relations with nature. The influential work of Rosemary R. Ruether offers an example of the nature of both ecotheology and ecofeminism. Reflecting on the radical implications of the Christian story of Jesus as an incarnation of God in the form of a servant, Ruether writes:

> In God's Kingdom the corrupting principles of domination and subjugation will be overcome. People will no longer model social relationships, or even relationships to God, after the sort of power that reduces others to servility. Rather they will discover a new kind of power, a power exercised through service, which empowers the disinherited and brings all to a new relationship of mutual enhancement.

The objective, explains Ruether, is a society governed by "equivalence and mutuality between men and women, between classes and races, between humanity and nature." Jesus' death on the cross signifies the death of the monarchical image of God, and it means an end to hierarchical and dualistic modes of thought. "A new God is being born in our hearts," writes Ruether, "to teach us to level the heavens and exalt the earth and create a new world without masters and slaves, rulers and subjects."[2]

As the awakening of American society to the greater earth community has deepened, increased attention has been given to Native American traditions. These traditions, like those of many tribal peoples throughout the world who have lived close to the land, reflect a keen sense of the interdependence of human culture and nature and involve a holistic ethic of respect for nature. Such "old ways" often contain a wisdom that Euro-Americans have forgotten, and a number of Native American tribal leaders and medicine people have emerged as important voices in the environmental movement. Among this group is Chief Oren Lyons of the Onondaga Nation and Iroquois Confederacy (the Six Nations). He gave this account of a Native American delegation's presentation before the United Nations in the 1970s.

> We went to Geneva—the Six Nations, and the great Lakota nation—as representatives of the indigenous people of the Western Hemisphere. We went to Geneva, and we spoke in the forum of the United Nations. For a short time we stood equal among the people and the nations of the world. And what was the message that we gave? There is a hue and cry for human rights—human rights, they said, for all people. And the indigenous people said: What of the rights of the natural world? Where is the seat for the buffalo or the eagle? Who is representing them here in this forum? Who is speaking for the waters of the earth? Who is speaking for the trees and the forests? Who is speaking for the fish—for the whales—for the beavers—for our children? *We* said: Given this opportunity to speak in this international forum, then it is our duty to say that we must stand for these people, and the natural world and its rights; and also for the generations to come.[3]

The voice of the Native Americans becomes in this way the voice of the Great Spirit, calling all peoples to reestablish a right relationship with the earth.

The international community began to focus on the environmental crisis in an organized fashion beginning with the United Nations Conference on the Human Environment held in Stockholm in 1972. Participants in the conference wrestled with the relationship of economic justice and environmental conservation, and increasingly discussions focused on the concept of sustainable development. The concept of development includes all that is involved in meeting human needs and improving the quality of life. Conservation is concerned with maintaining the resources, or natural capital, necessary to development. When conservation and development are in harmony, there is *sustainable* development. Strategies of sustainability respect the wisdom in the Native American proverb: "The frog does not drink up the pond in which he lives." In 1980 the International Union for Conservation of Nature and Natural Resources (IUCN), a union of over 450 governments and nongovernmental organizations, released the first World Conservation Strategy. It had been developed through joint efforts with the United Nations Environment Programme (UNEP), UNESCO, the Food and Agricultural Organization of the UN, and the World Wildlife Fund, and it called for "a new ethic, embracing plants and animals as well as people," an ethic that would ensure sustainable development.[4]

The first major international effort to formulate a global ethic of sustainable development was led by the President of Zaire, Mobutu Sese Seko, and resulted in the United Nations *World Charter for Nature*.[5] It was overwhelmingly approved in 1982 by the UN General Assembly (111 to 1) with the United States casting the sole negative vote. What is especially noteworthy about the UN *World Charter for Nature* is that it is a statement that affirms biocentric as well as anthropocentric values. The Charter is based on the assumption that "mankind is a part of nature and life depends on the uninterrupted functioning of natural systems." In addition, it declares that "every form of life is unique, warranting respect regardless of its worth to men, and, to accord other organisms such recognition, man must be guided by a moral code of action." The text of the *World Charter for Nature* is printed in the appendix to this volume.

Building upon this development, in 1983 the United Nations organized a World Commission on Environment and Development to develop "a global agenda for change" that would achieve sustainable development by the year 2000 and beyond. Appointed to chair the commission was the Norwegian Prime Minister Gro Harlem Brundtland, the only political leader who had served as an environment minister before becoming prime minister. The Brundtland Commission report was issued in 1987, under the title *Our Common Future*, and was quickly translated into seventeen different languages. Its analyses of the problem and its proposals for sustainable development set the new framework for international debate and planning.[6]

Secretary General of the Brundtland Commission Jim MacNeill summarized the findings of the report: "Much of what God created, man is now destroying. . . . The world's economic and political constitutions are seriously out of step with the workings of nature. . . . Our economic and ecological systems are now totally interlocked in the real world, but they remain almost totally divorced in our institutions."[7] The answer, argues MacNeill, is a worldwide integration of environmental considerations with economic and political decision-making processes. "Sustainability should become the test of sound economic policy." Explaining the conclusions of her World Commission report, Prime Minister Brundtland commented: "Fundamentally, 'sustainable development' is a notion of discipline. It means humanity must insure that meeting present needs does not compromise the ability of future generations to meet their needs. And that means disciplining our current consumption. This sense of 'intergenerational responsibility' is a new political principle, a virtue, that must now guide economic growth."[8] Brundtland recognized the religious and ethical dimensions of the problem:

> A new cultural ethos is the main thing. That ethos, I believe, is intergenerational responsibility. If that ethos is not accepted almost as a religious belief, we cannot convince anyone that we must change the way we live. If we cannot make people realize that living as we do will make it impossible for their grandchildren to live at all, they won't change. If people believe this is true, it is a premise that can reach both minds and hearts.[9]

The Brundtland Commission report also stressed that "it is impossible to separate economic development issues from environment issues." It found that "poverty is a major cause and effect of global environmental problems." In short, problems of economic and social justice are interrelated with issues of environmental ethics.

When IUCN in collaboration with UNEP and the World Wide Fund for Nature set out to prepare a second version of the World Conservation Strategy, building on the Brundtland Commission report, it made a decision to put special emphasis on the need for a "world ethic of sustainability" and the transformation of human attitudes and values. The senior consultant to the second World Conservation Strategy project, Robert Prescott-Allen, explained that: "Sustainability calls for a fundamental transformation in how people behave. Changes in behavior can be assisted by laws and incentives, and by material help where needed. But the changes will last only if they arise from attitudes and practices based on a commitment to sustainability."[10] Toward the goal of formulating a world ethic of sustainability a special IUCN international Ethics Working Group was formed in 1987 under the leadership of J. Ronald Engel. For Engel and his committee members, sustainable development is a name for "a new morality as well as a new economic strategy." It is "a new moral conception of world order." Explaining further, Engel writes:

> "Sustainable," by definition, means not only indefinitely prolonged, but nourishing, as the Earth is nourishing to life, and as a healthy natural environment is nourishing for the self-actualizing of persons and communities. The word "development" need not be restricted to economic activity, . . . but can mean the evolution, unfolding, growth, and fulfillment of any and all aspects of life. Thus "sustainable development," in the broadest sense, may be defined as *the kind of human activity that nourishes and perpetuates the historical fulfillment of the whole community of life on Earth*.[11]

The Brundtland Commission report had been written primarily from an anthropocentric perspective, but the World Charter for Nature and the World Conservation Strategy of IUCN go beyond that, reflecting to some degree the spirit of Albert Schweitzer's "Reverence for Life" and the biocentric and holistic perspective of the American environmental writer Aldo Leopold.

Formal publication of the second World Conservation Strategy occurred in 1991, under the title *Caring for the Earth*. A draft of this document was printed and distributed to concerned groups in the fall of 1990 in preparation for IUCN's World Assembly meeting in Australia at the end of the year. It contained a brief statement outlining "the elements of a world ethic of sustainability." This world ethic focuses both on relations among people and on human relations with nature, recognizing that the liberation of nature and the liberation of people are fatefully intertwined. A revised version of this ethic of sustainability is printed in chapter 7 with the editors' introduction to Robert Prescott-Allen's essay. Here one finds the elements of a new planetary vision that joins democracy, ecology, and moral faith. Some animal rights activists and deep ecologists would argue that it does not go far enough, and the debate will continue, but it effectively focuses the discussion on the critical issues. The next major forum for international deliberation will be the United Nations Conference on Environment and Development scheduled for June of 1992 in Rio de Janeiro, Brazil.

The global discussion of environmental issues has revealed that there are many diverse cultural pathways that lead to a faith in an ethics consistent with ecology and social liberation. As this discussion has developed, it has increasingly included cross-cultural and interfaith dialogue. For example, many Americans turned to Eastern religious traditions like Taoism and Zen in the sixties and seventies, in part because these traditions taught that there are religious and ethical values to be realized in relationship with nature. In the mid-1980s western environmental philosophers began to engage seriously in comparative philosophical studies involving diverse Eastern and various Western approaches.[12] A global environmental philosophy movement is now emerging.

During the late seventies and the eighties the leadership of the world's religions began to address the challenge posed by the environmental crisis. The first interfaith dialogue involving prominent representatives from the five world religions occurred in Assisi, Italy, in 1986. The gathering was sponsored by the World Wildlife Fund. "The Assisi Declarations" that were issued following this historic event did not contain new ethical or philosophical ideas, but they did mark the beginnings of interfaith cooperation complementing the international collaborative work of philosophers, scientists, activists, and government leaders. Two years later Pope John Paul II and the Dalai Lama met in Rome to discuss issues of "world peace, spiritual values and protection of the earth's natural environment."[13] Also, in 1988 leaders from the world's religions gathered together with statesmen in Christ Church College at Oxford University for a Global Conference of Spiritual and Parliamentary Leaders on Human Survival to explore strategies for collaboration in the pursuit of peace and planetary well-being.

A second Global Conference followed in Moscow early in 1990. Attracting several hundred religious leaders as well as scientists, educators, and political leaders from eighty-three nations, it was focused on issues of environment and development. Keynote addresses were given by Prime Minister Gro Brundtland and Native Elder Audrey Shenandoah of the Onondaga Nation. President Mikhail Gorbachev addressed the gathering, speaking of a new "unconditional code of ecological ethics." These international undertakings are being complemented by projects such as the North American Coalition on Religion and Ecology (NACRE), an interfaith organization working in collaboration with thirty environmental groups in an effort to promote the cause of "Caring for Creation" among diverse religious communities.[14] In 1993 representatives of the world's religions will gather for historic meetings in Bangalore, India, and Chicago in celebration of the centenary of the first World's Parliament of Religions held in Chicago in 1893, and the issue of environmental ethics will be high on the agenda.[15]

The Middlebury College symposium on "Spirit and Nature: Religion, Ethics, and Environmental Crisis," was organized in the Assisi spirit of interfaith dialogue and cooperation. The principal speakers were Audrey Shenandoah, a Native Elder of the Eel Clan of the Onondaga Nation; Ismar Schorsch, Chancellor of the Jewish Theological Seminary in New York; Seyyed Hossein Nasr, Professor of Islamic Studies at George Washington University; Sallie McFague, Professor of Theology at the Divinity

School of Vanderbilt University; J. Ronald Engel, Professor of Social Ethics at Meadville/Lombard Theological School; and Tenzin Gyatso, His Holiness the 14th Dalai Lama, representing Tibetan Buddhism. All of the speakers contributed to the global environmental perspective by drawing on and interpreting the wisdom of their own traditions. Other symposium events represented in this volume include Robert Prescott-Allen's report on "Caring for the World" and a panel discussion among the various speakers. An essay and an epilogue by the editors seek to put the issues addressed by the symposium in a larger historical context and reflect upon the meaning of the event. A selected bibliography is included for readers who want to pursue these issues further.

An international art exhibition was an important part of the symposium, and a number of images from it are included in this volume. The intent of the exhibition, which involved thirty-six objects from fifteen different cultural traditions around the world, was to focus attention on the interrelatedness of human civilization and nature and to celebrate the beauty and awesome mystery of life in the natural world. The gallery as a whole was conceived of as a Tree of Life. A giant circular redwood table by George Nakashima was set in the center of the gallery representing the trunk of the Tree of Life and the other works in the show were viewed as being contained within the branches of this cosmic tree. The Tree of Life appears in many different cultures as a symbol of the sacredness of creation, the bonds between the spiritual and the natural, the continuation of life, and peace. As images of beauty and sources of oxygen, food, shelter, and building materials, trees are wonderful expressions of the interconnectedness of culture and nature. The symbol of the Tree of Life is, therefore, an especially meaningful image for the late twentieth century. In addition the wholesale destruction of forests in both the northern and southern hemispheres reflects the ecological breakdown of the entire biosphere and the spiritual crisis in humanity's relations with nature. In the exhibition the Tree of Life image appeared on a Torah Mantle and an Islamic prayer rug, and it was present in the form of the Christian Cross and the Great Tree of Peace from the Iroquois tradition. There was also in the exhibition a Tree of Death sculpted out of the charred remains of trees from a destroyed Brazilian rain forest. A stark reminder of the ecological-spiritual crisis of our time, the Tree of Death stood at the opposite end of the gallery from the NASA photograph of Planet Earth, a symbol of the beauty, community, and peace that is the threatened promise of life on earth.[16]

NOTES

1. Copies of the documentary, entitled "Moyers / Spirit and Nature," may be acquired by writing: Spirit and Nature, Box 2284, South Burlington, VT 05407, or by calling 800-336-1917. Transcripts are also available; write: Journal Graphics, 267 Broadway, New York, NY 10007, or call 212-227-READ.
2. Rosemary R. Ruether, *Sexism and God-Talk: Toward a Feminist Theology* (Boston: Beacon Press, 1983), 11, 22, 30.
3. Oren Lyons, "Our Mother Earth," in D. M. Dooling and Paul Jordan-Smith, eds., *I Become Part of It: Sacred Dimensions in Native American Life* (New York: Parabola Books, 1989), 273.
4. J. Ronald Engel, "Introduction: The Ethics of Sustainable Development," in J. Ronald Engel and Joan Gibb Engel, eds., *Ethics of Environment and Development: Global Challenge and International Response* (London: Belhaven Press, 1990; rpt. Tucson: University of Arizona Press, 1990), 2–3.
5. Ibid.
6. Gro Harlem Brundtland, "Chairman's Foreword," in World Commission on Environment and Development, *Our Common Future* (New York: Oxford University Press, 1987), ix–xv.
7. Jim MacNeill, "Sustainable Development: Getting through the 21st Century" (Address delivered to the John D. Rockefeller, 150th Anniversary Conference on Philanthropy in the 21st Century, Pocantico Hills, N.Y., 28 October 1989).
8. Gro Harlem Brundtland, "The Test of Our Civilization," in *New Perspectives Quarterly* 6, no. 1 (Spring 1989): 5.
9. Ibid., 7.
10. Robert Prescott-Allen, "Caring for the World" (Address delivered at the Symposium on "Spirit and Nature: Religion, Ethics, and Environmental Crisis," Middlebury College, Middlebury, Vt., 14 September 1990).
11. Engel, *Ethics of Environment*, 10–11.
12. See, for example, J. Baird Callicott and Roger T. Ames, eds., *Nature in Asian Traditions of Thought: Essays in Environmental Philosophy* (Albany: State University of New York Press, 1989), xi, xiii-xxi, 1–21.
13. Statement issued in Rome on 14 June 1988, by His Holiness the Pope, John Paul II, and His Holiness the 14th Dalai Lama.

14. North American Coalition on Religion and Ecology, Donald B. Conroy, Director, 5 Thomas Circle NW, Washington, D.C. 20005.
15. The centennial observance in India is being coordinated by four interreligious, international organizations which seek to further the spirit of the 1893 World's Parliament of Religions: the International Association for Religious Freedom (IARF), the World Congress of Faiths (WCF), the Temple of Understanding (ToU), and the World Conference on Religion and Peace (WCRP). The centennial observance in Chicago is being coordinated by the Council for a Parliament of the World's Religions, Daniel Gomez-Ibañez, Chairman, P. O. Box 377760, Chicago, Ill. 60637–9998.
16. For a catalogue of the Middlebury College "Spirit and Nature" art exhibition, see Steven C. Rockefeller and John C. Elder, eds. *Spirit and Nature: Visions of Interdependence* (Middlebury: The Christian A. Johnson Memorial Gallery, 1990).

CHAPTER 1

A Tradition of Thanksgiving

Audrey Shenandoah

Fig. 1. According to the oral traditions of the Iroquois Confederacy, as well as of some Asian peoples, the earth rests on the back of an ancient turtle that swims upon primal waters. Rising from Turtle Island in Arnold Jacobs's (1942–) Creation *(1983) is the Great Tree of Peace, which symbolizes the formation of the six-nation Iroquois Confederacy and the law of harmony among all beings. The eagle watching from the top of the great pine tree is the guardian of peace. (Private collection.)*

One of the principal Native American traditions is that of the Haudenosaunee or Iroquois Confederacy, which includes the Mohawk, Oneida, Onondaga, Cayuga, Seneca, and Tuscarora nations. Elder Audrey Shenandoah is a Clan Mother of the Onondaga nation and has devoted much of her life to preserving and sharing the sacred teachings that lie at the heart of traditional Iroquois culture. The Haudenosaunee believe that if society at large is to address the environmental crisis effectively, it must first undergo a spiritual transformation and embrace the spiritual principles and attitudes associated with these teachings.[1]

The spiritual values that lie at the heart of the common life of the Haudenosaunee are well symbolized by the Great Tree of Peace, which is envisioned as an evergreen tree with a timeless message of cosmic harmony. Its great roots stretch out to guide those from the four directions who would join with the Iroquois in the pursuit of peace. Its trunk is straight, reaching to the sky world. The Great Tree of Peace signifies the Great Law, which emphasizes the principles of equity, justice, and peace. Closely associated with the Great Law are respect for the earth and a profound sense that humanity is a part of a great family which includes all of nature. Out of this sense of respect for and belonging to nature grows a deep sense of thanksgiving for all the beings and forces in nature that sustain life. The Great Tree of Peace stands, too, as a reminder to the people of their power when they are of one mind in devotion to what is true and good.

Audrey Shenandoah's keynote presentation at the "Spirit and Nature" symposium opened with a traditional thanksgiving prayer. Her remarks offered a meditation on the meaning of this prayer, and her manner of speaking reflected the Onondaga oral tradition, in which interwoven repetition serves to emphasize basic themes. Ceremonies and meetings among the Haudenosaunee always begin with the thanksgiving prayer that Shenandoah used to open the symposium, but it may not be recorded in film or printed and, accordingly, does not appear here. With her permission, however, we offer a brief summary.

The prayer begins by acknowledging the presence of all those gathered together and by giving thanks for the opportunity to meet. It then proceeds to name and acknowledge the various elements of the creation that sustain the diverse community of life, beginning with our mother, the Earth, and including the waters, plants, animals, fish, birds, thunderclouds, stars, and finally our elder brother, the Sun, and our grandmother, the Moon. No one species is superior, no part of nature is unimportant. At each step in the cycle of expanding awareness generated by the prayer, the leader returns to a refrain, calling upon those assembled to put their minds together and as one mind to give acknowledgment and a great thanksgiving for the part of creation being named. The prayer concludes with a thanksgiving addressed to the Creator.

THIS is the thanksgiving address that our people have every time we get together, no matter how many or how few. And at every ceremony this is done. It is done at the beginning, and it is done at the closing.

You notice the very first component of this address is people. So people must be very important in the eyes of our Creator, if we should first give thanksgiving to each other and for each other. But all of the elements within this cycle of creation work together harmoniously, in balance, and our place is to keep the balance, to use them right, to use them wisely, and to remember to give thanksgiving.

Many of the elements are now out of balance, and we have forgotten in many instances to give the proper thanksgiving. Water is life. Water is very important to all living things. Without clean water there will be no life. Money should not be asked for water, for it is a gift given freely from our Creator. Technology and industry I am sure could find the ways to purify the water that has been polluted and abused. Technology and industry I am sure could do much research and find out how to correct quite a bit of the imbalance that is happening in our world today.

We are responsible for seven generations in my tradition, seven generations into the future. Our leadership must not make decisions that are going to bring pain or harm or suffering seven generations in the future. As individuals, we have the same mandate. This is a cause for great concern among my people, for we live in a society that feels dominant over the rest of the elements of the creation. But we are not. All the rest of the cycle could continue, if we were to drop out. It would probably continue very well, and the earth could heal itself. Other creatures would continue in their work. But if any single one of the rest should drop out from the cycle, we would be crying, we would be suffering, and our children would be suffering.

It is one of the main thoughts within the Thanksgiving Prayer that there be much repetition. It is purposely repeated over and over again, that we must put our minds together, and that we must give a thanksgiving for all of life, for all that gives us sustenance here on this earth. So much repetition allows for one to absorb the thoughts and in that way they become a part of one's being.

There are many, many people on the earth today who are poor and hungry. There are also many, many stockpiles of material wealth, people in business only for making money and not caring about the harm that may come to the people, the consumers of their products. They do not care whether what they are selling is healthful or beneficial, or what it is going to be used for in the future.

This is one of the main components of our thanksgiving address, that we must give thanks, and remember that we are responsible for seven generations into the future. I think this is a thought well worth looking into by all of industry and technology. But I ask, and we ask, will they, are they capable of doing this? Are they capable or willing to use a Good Mind, humane thought, as they continue to be producers of goods.

We cannot wait for multinationals, we cannot wait for huge programs to do all of this. It is up to each and every individual, once you have heard the message. Then you have the responsibility to teach it. This is the philosophy of the tradition that I come from. That is why there is so much repetition, so that you cannot say I do not know that, I did not hear that.

In today's times, we have a very difficult situation. Many of our young people no longer speak and understand our native language. And so we have to do a very difficult task, like today, putting this message into a foreign language, the English language. Our own language is still connected to the land. But in the English language, you have to use a whole bunch of cold words to try to describe what in any native language, because it's a land language, can be said completely, rounded out and whole.

So we have a difficult time giving our young people this understanding, and at the same time trying to teach the language. Because like any other language, there is the language of the ceremony, and there is the everyday language that you use as you move about with your friends. And so we have this very, very difficult time, not only among my own people, who are the Haudenosaunee, but among our native brothers all over this Turtle Island.

And I think that indigenous peoples all over the world have this basic principle of a respect, a deep love for the land, a very real connection, a relationship to all of those elements that are mentioned in our thanksgiving. I believe somehow we have to try to reach the people in power, to try to get them to listen to the indigenous people, no matter where they are.

We had a meeting with some very interesting spiritual people this August in my own territory. There were indigenous peoples from several parts of the world, and the message was always the same. We were not surprised, because we have met before with those from other native lands, people who still feel related to the land and the components of the cycle of life. The message is the same, that we should respect our Mother, the Earth, and all of the components of the cycle of creation and that we should not continue the abuse that has been going on for so many years. Much technology can be used in a good way, but it has to be thought over. It has to be looked at. Proper changes and alternatives need to be acted upon in order to ensure well-being and peace of mind for future generations.

There is a spiritual foundation that all of us were born with. It is inherent to every kind of people, a spiritual connection to this earth, to this land. This word "culture" is very much overused, I say. It is our way of life, and it is our connection to all of the rest of creation. That is our way.

We dance, we sing, expressing thanksgiving, expressing good feelings, strong spirits. We are misunderstood by many, many people. We have been portrayed as people who worship the sun, who worship elements of the creation, and this is incorrect. We give thanks, we respect, and we acknowledge *all* the rest of the creation. And this is what our ceremonies are for. This is what our ceremonies are all about.

They take up a whole lot of our time. We just finished one ceremony that lasted six days. And we will have another coming up very soon. We have a thanksgiving celebration for everything in its season. We also have a thanksgiving celebration for what many people are afraid to think about and to approach, even though it is a very important part of our life. And that is what in English is called death. We have a celebration for a person passing from this world into the other. And it has equal stature in our minds to any of the other of our thanksgiving celebrations.

We do not shelter our little ones from that thing that is called death, and so feared by so many people. It is a part of our life; it is a part of our lives. And so our little children are not shocked and frightened and driven into all kinds of emotional disturbances when death comes close to them. And it is right in the speeches, it is right in the ceremony, that we should instruct our children from the time of reason in thanksgiving, respect, acknowledgement, and also in what is called passing from one world to the other. And so it is really a complete cycle, if we will take the time to look at it. This passing comes in its time. We do not have to know the time, but it comes in its time. Our prophecies tell us of a time when the principal ones of all these elements will face destruction.

As we flew over the mountains from New York to Vermont, it was pretty cloudy. But I watched—because I've never seen the Adirondacks and the maple forests from the air, coming in this direction. Our prophecies from a long, long time ago tell us that this Chief of the Woodlands, the Maple, will begin dying from the top. I think a lot of people have heard this, since it began happening.

We were also told a long, long time ago that things would happen to the animal life. This was before the invasion of the foreigners to this land, and while things were still what you call "a wilderness." In our word there is no such thing as a wilderness. It is only a free place. It was not feared, and there is no need to fear it now. People no longer know what is in that one-time free wilderness. So many things that we have to be on guard about.

So I ask that everyone here have this thanksgiving in our minds. Somehow we have to get the message, as I said before, to those who are in command of the technology and the industry. We have to reach and soften their thinking if it is possible, letting as many people as we can know that we must make a turnabout now. *Dane'tho.*

Following Elder Shenandoah's talk, she responded to questions from the audience. Several of those questions and answers follow.

Q: Could you please say something about how children are given guidance in your culture so that they grow up with this love and reverence for life?

SHENANDOAH: Within our longhouses, and within our ceremonies, I know of only two ceremonies where children are not allowed. Our children are brought with their mothers when they are new babies to get their names, for one, and also to do our sacred feather dance for the first time. And so the children are brought from the time they are infants to the longhouse, by their mothers, or their grandmothers, or their aunts.

Many of the young parents do not know the language. So we have to spend time with these young parents, so that they can in turn instruct their children. I guess this makes our job three times harder.

We have an extended family within our longhouse tradition. We are not like the nuclear family—mother, father, children, perhaps grandparents. We still carry on the tradition of clanship. Clan families meet, and the traditions are passed on at these times. During some of our ceremonies, we have to meet at people's houses in clans before we go to the longhouse. This is another time that traditions are passed on.

Then there are the storytelling times. They are not held as often as they used to be when I was a little girl, but we still have them. Children learn by listening to the elders and to their teachers in school.

Q: Please tell us a little more about your people's vision of the other world, and perhaps the word for death.

SHENANDOAH: Our Creator is a loving creator. Our Creator is not one who is vengeful, or who is going to punish. We punish ourselves if we do not follow the instructions. We bring punishment upon ourselves here on this earth. Our ways teach us that we should try in this time on earth not to have any disharmony within our own space. Each person has his own space that he is responsible for. Your responsibility is to keep peace within that space around you, within your own space. You know this would do a great deal toward world peace today, if we all knew that, if we all practiced that.

When we pass from this earth, we pass into what is called the spirit world. And there, all of the ceremonies that we do at our longhouses are also done. We are told that the minute a person takes his last breath here on this earth, a singer there, whoever it is, beats down on the sacred turtle rattle, and the ceremony begins.

So if you are faithful to ceremonies, if you are faithful to the teachings of our ways, these are the things that you will have around you in the spirit world, around you all the time. And whatever your way is here on this earth, that is the way that you will be there unless you find yourself and correct yourself. But only you can correct yourself.

Q: I have two questions. You mentioned the prophecy that the maples would begin dying from the top down. I wonder if the prophecy went further and spoke to a remedy for that situation? And second, I wondered if there was a word for religion in your language?

SHENANDOAH: There is a word for what they call Christianity. Otherwise, no. I guess because it is so much a part of the people to live this way that it was not considered a religion. It is just the way to live.

And the maple tree. No, not a physical remedy. Except that we should do our ceremonies. We have a ceremony for healing, which we do. But at the very end of one of our teachings, it says it is up to the people how long this all will continue, how long this will be. How long we will have the maple is up to the people. How long we will have the strawberry is up to the people. How long we will have water to survive is up to the people. Every individual *can* make a difference!

NOTES

1. Arnold Jacobs, "Creation," as cited in *Iroquois Arts: A Directory of a People and their Work* (Warnerville, N.Y.: Association for the Advancement of Native North American Arts and Crafts, 1983), 405. See "Interview with Oren Lyons, Onondaga Chief," in Andie Tucher, ed., *Bill Moyers—A World of Ideas: Public Opinion from Private Citizens* (New York: Doubleday, 1990), 2:175–86.

CHAPTER 2

Learning to Live with Less: A Jewish Perspective

Ismar Schorsch

Fig. 2. On this Eastern European Torah Mantle (1870–71), the Tree of Life symbolizes the hope for messianic redemption. The birds are associated with the protection of the Crown of Torah. The lion and the deer reflect a saying of Rabbi Judah ben Tema: "Be bold as a leopard, light as an eagle, swift as a deer, and strong as a lion, to do the will of your Father who is in Heaven" (Pirke Avos, 5:23). (Jewish Museum Art Resource, New York. Gift of Dr. Harry G. Friedman, F3875.)

Over the past twenty years, the subject of Judaism and ecology has gradually emerged as a significant issue for Jewish religious thinkers. Dr. Ismar Schorsch is one of those religious leaders and scholars who in recent years has brought the issue to the fore. Initially, Jewish as well as Christian theologians concerned about the environment found themselves responding to the attacks mounted by Lynn White, Jr., an environmentally concerned historian who argued, with other critics, that the Judeo-Christian tradition has involved a dangerous anthropocentrism and a problematic dualism of the spiritual and the natural. Some Jewish scholars, however, have found grounds in their tradition for a more constructive approach to the relation between religion and ecology.

For example, they have identified in the Hebrew Scriptures and the Talmud, or in the mystical thought associated with the Kabbalah and Hasidism, or in the philosophy of a figure like Martin Buber, various ways of developing a Jewish theological understanding that supports respect for nature and a concern for environmental ethics. In his explorations of Judaism and ecology, Ismar Schorsch has taken this kind of position. He endeavors to demonstrate that when thoughtfully interpreted and related to the current situation, the sacred traditions of historical Judaism include the theological vision and moral wisdom needed to address the spiritual problems that underlie the environmental crisis.

In his essay Schorsch offers a Jewish perspective on the need for a new sense of global community and responsibility. His central question is: "Can religion responsibly imbue the individual citizen with a spiritual renewal that will ennoble his worldview, enlarge his inner life, and temper his wants?" In the biblical understanding of God and human nature, the law set forth in the Bible and Talmud, and the Jewish love of learning, he finds a positive answer to this question. These Jewish traditions affirm "a culture of self-restraint" that respects nature, suggesting how a people aided by faith can overcome the consumerism and materialism that characterize our environmentally destructive way of life.

I

LAST September (1989), the Jewish Theological Seminary of America devoted its annual High Holiday message, published as a full-page ad in the *New York Times* and elsewhere across the land, to the environmental crisis. The public reaction to that blunt religious alert was, with few exceptions, highly favorable and exceeded in scope anything the seminary had experienced in nearly forty years of annual messages. The response provides yet another small confirmation of the welcome development that, since the first "Earth Day" two decades ago, consciousness of the cumulative damage inflicted by human society (industrial and impoverished) on the planet has become acute and global.

But the magnitude of the problem transcends national boundaries, for only international cooperation can reverse the rise in acid rain, the depletion of the ozone layer, the destruction of the rain forests, the pollution of the Danube, or the exponential growth of humankind. In this regard, nothing could be more auspicious than the dramatic end to the cold war, which surely removes the immediate threat of a nuclear nightmare and should release eventually vast national resources for more constructive purposes. The ecological havoc wrought by communist regimes hell-bent on heavy industry and military hardware has yet to be fully assessed, but for the moment at least it was most aptly symbolized by a photograph last winter in the Sunday Magazine of the *New York Times*—a hospital for respiratory diseases somewhere in the coal regions around Krakow, placed in an excavated mine because the air down under is cleaner than the air above ground. With expanses of eastern Europe bordering on the surrealistic, a reunited continent is beginning to evince the political will to confront its common ecological destiny.

No less encouraging is the growing alliance between scientific and religious leaders on environmental issues. The two communities have been at loggerheads for most of the modern era, most recently over the ludicrous proposition to teach creationism in the public schools as a valid scientific alternative to evolution. Indeed, in the seminal essay of 1967 by Professor Lynn White, Jr., writing in the shadow of Max Weber, the contemporary environmental movement was itself launched by a spirited attack against the biblical "hubris" that purportedly shaped the domineering mind-set of the Christian West.[1] But of late, a spirit of collaboration has replaced the instinct for combat. Religious leaders have begun to heed the warnings of concerned scientists about the deteriorating state of the planet, with some 270 of them from around the world prepared to identify themselves with the "Appeal for Joint Commitment in Science and Religion" issued by a group of twenty-three renowned scientists from six countries at the Global Forum of Spiritual and Parliamentary Leaders held last January (1990) in Moscow. The forum, in fact, unveiled a glimpse of both hopeful trends: the converging of international and disciplinary lines on the welfare of the earth.

What is it that religion might contribute? In the West, where the environmental crisis springs from prosperity, and not poverty, the ultimate solution will need to go far beyond the corridors of Washington or the headquarters of corporate America. For, in our society, the root of this macro problem is entangled with the values of the average American.

I will give but one graphic example. There is surely a link between the unnerving fact that the United States, with but 2 percent of the world's population, annually consumes one quarter of the world's oil production and the vast parking lot filled to overflowing at the regional high school in south Jersey where I occasionally play tennis. The sight of this parking lot always reminds me of the senior admitted to UCLA who says: "I would have gone, but I couldn't get a parking space." I have yet to see a bicycle rack at that school. Clearly the issue here is not one of transportation, which is readily available in other forms—a school bus, a bike, or, heaven forfend, even walking—but a state of mind that venerates the car as the chic mode of travel. And parents, enamored in the same web of materialistic values, indulge their children and end up exposing them to untold risks. In short, I heartily concur with Wendell Berry, when he writes: "Our environmental problems . . . are not, at root, political; they are cultural. . . . Our country is not being destroyed by bad politics; it is being destroyed by a bad way of life."[2]

To attack the problem, then, from the bottom up, at the human level, is the challenge of religion. In the United States, we are about to endure the closing of another frontier—the enthralling vista of a world of limitless bounty. Can religion responsibly imbue the individual citizen with a spiritual renewal that will ennoble his worldview, enlarge his inner life, and temper his wants? Obviously, that question cannot be answered generically; each religious community will have to explore the treasured resources of its own tradition with a view to enriching the life of both its members and society as a whole. For my own part, I wish to offer a portrait of Judaism as a millennial effort to foster a religious culture of self-restraint that intuitively respects the value and integrity of its natural environment. My argument is essentially systemic and structural and not overly concerned about harmonizing the details of Judaism with our modern sensibility and scientific knowledge, though I deem them on the whole to be informed with deep wisdom. Nor shall I try to make of it a natural religion with a special capacity for communing with nature. Rather, I believe Judaism to have constructed a sacred canopy of legal regulations, theological notions, and intellectual values on a bedrock of a transcendent God concept, the outcome of which is a decidedly modest sense of man's place and purpose in the universe.

II

Let me begin my partial phenomenology of Judaism with its pervasive legal character, long regarded as the converse of spirituality. In Judaism, freedom is a blessing that must be harnessed, structured, and channeled to become beneficent. In the biblical narrative, the revelation at Sinai follows directly upon the redemption from slavery so that freedom might be yoked to a grander vision and translated into a system of deeds. God's commandments—concrete and multitudinous—create the building blocks for religious community and the spiritual notes for individual expression as the rhythm and rites of the sacred lessen the burden of the mundane.

Broadly speaking, Jewish law, as enunciated in the Hebrew Bible and expanded or curtailed in the Talmud, governs how one relates to God and to fellow human beings. More to the point, within each category it often comes to regulate a Jew's relationship to the natural world. The product largely of an agrarian society, *halakha*, a word that etymologically connotes boundary, is rife with regulations designed to rein in man's unlimited use of his environment.[3] What is consistently deemed acceptable is far short of what is within human reach. The answer to the vexing question of "how much is enough" is not determined solely by commercial interests.

My examples are drawn from three different areas of life. Ideally, Scripture would have us live as vegetarians. The permission to eat meat is not granted until after the flood, and then only with a visceral prohibition against the consumption of blood.[4] After Sinai, the Bible restricts that privilege to ruminant animals with cloven hoofs and fish with fins and scales.[5] The later rabbinic prescriptions for ritual slaughtering appear to be an echo of an early stage in Israelite history when all slaughtering was connected to the cult, and yet another reminder of the sober fact that the sustaining of human life comes at the expense of other life.[6]

In this spirit, Jewish law also developed a sensitive set of injunctions against the inflicting of pain on animals. We are bidden not to muzzle an ox while threshing, or yoke together animals of different sizes (such as an ass with an ox) for plowing, or kill on one and the same day a cow and its calf.[7] Similarly, when coming upon a bird's nest with eggs or fledglings still cared for by the mother, we are to release the mother.[8] Compassion for animals is also what seems to prompt the Bible to repeat in three different passages the prohibition against boiling a kid in the milk of its mother, which becomes the basis for the later talmudic separation of the consumption of meat and milk products.[9] Indeed, the Bible even delivers the admonition not to extend our animosity to the animals of our enemy. "When you see the ass of your enemy prostrate under its burden and would refrain from raising it, you must nevertheless raise it with him."[10] Culminating that deep-seated concern is the prophetic inclusion of animal life in the messianic vision, though no one quite matched Isaiah's imagery of the restoration of pristine harmony with man and beast bonded once again in lasting fellowship, a prophecy made part of America's cultural heritage by Edward Hicks's primitive painting in 1848 of *The Peaceable Kingdom*.[11]

The use of land is similarly restricted by religious values of some prudential consequence. Perhaps best known is the institution of the sabbatical year in which the land was to lie fallow and whatever grew naturally was to be shared by man and beast.[12] But the Bible also enjoins against harvesting the corners of your field or returning for the gleanings—both to be left to the indigent without offense to their dignity.[13] Crops were not to be mixed within the same field nor animals of different species interbred, an apparent expression of respect for the divine ordering of animate things.[14] In fact, God was perceived to be an active partner in the production of all life, and hence first fruits and offspring, even human, belonged to God, to be given to the priests of His sanctuary or in the case of the first born male child to be symbolically redeemed.[15] Also noteworthy are the proscription to leave untouched the fruit of any new fruit tree for the first four years,[16] and the admonition not to destroy fruit trees in time of war, which later Jewish law expanded to include the wanton destruction of any useful object or resource.[17] Even the bearing of children, hardly unrelated to our demands on the land, is guided by the same spirit of moderate restraint. According to the early rabbis, the commandment "to be fruitful and multiply" was deemed fulfilled as soon as one had given birth to either two sons (the school of Shammai) or a son and a daughter (the school of Hillel).[18]

Lastly, an unmistakable strain of self-denial runs through the Jewish calendar. From the sacrificial cult of the temple to the synagogue of rabbinic Judaism, it is the absolute cessation of work that distinguishes the celebration of Jewish holy days. Whether it be the annual harvest festivals of Passover, Shavuot, and Succot or the weekly sabbath day, spiritual renewal is effected through physical contraction. On two introspective occasions of the year, the self-discipline extends to a twenty-four-hour fast—on Tishe B'Av to recall the national calamities of Jewish history and on

Yom Kippur for each individual Jew to repair his or her own relationship with God.[19] The regimen of rest is meant to restrict our strength as much as to restore it, to deflate our arrogance as much as to ennoble our spirit. The final intent of the opening chapter of Genesis is to pave the way for limits. It anchors the later and unnatural command to rest weekly in a cosmic act and induces us to acknowledge in deed, not word, that our dominion is only partial. To spend one-seventh of one's life in "unproductive" rest is scarcely a mark of absolute power.

The reference to Genesis brings me to the theological worldview which undergirds this self-limited praxis of Judaism. The Bible as it stands most assuredly starts twice, with two different accounts of creation that turn on two distinct readings of human nature, a contrast worth pondering. In the first chapter, we are treated to a gender-neutral portrait of Adam One as a superior man created in God's image and wedded to a wife of equal stature. With the whole world for their domain, they are blessed together by God and given the mandate to bring the earth and all its inhabitants under their sway. No blemish mars their perfection; no restraint curbs their power. God's creative activity culminates in the flawless formation of both, and indeed "there is nothing to be seen more wonderful than man."[20]

Not so the Adam Two of the second chapter. Remarkably, the Bible felt compelled to add a more sober verdict on human nature. The fragile, deficient, and limited man of Eden approximates the ordinary mortal of our everyday experience. Not only does the story not claim that he resembles God, it stresses that the substance of his being was taken from the dust of the earth. God created Adam Two without a mate, set him down in a circumscribed spot "to till it and tend it," and saddled him with at least one prohibition. Nor is any blessing bestowed to exalt his humanity. His mission, accordingly, is not to conquer the earth but to steward his garden. And as the subsequent chapters of Genesis dramatically reveal, he fails quickly and miserably at even this meager task. Adam Two, and not Adam One, is the subject of human history, a creature whose passions repeatedly overwhelm his intelligence and abuse his freedom to act. The exalted opening chapter of Genesis may project a tantalizing glimpse of human nature as just a "little less than divine," but it fails abysmally to illumine its daily dark, demonic side.[21] Much like an old stereoscope, the two perspectives are placed side by side to coalesce into a full-blooded, three-dimensional portrait of humankind.

By now it should be clear that the religion of the Bible is rooted in the conception of human nature as personified by Adam Two. The stasis of creation is shattered by man and his capacity for evil without end. What drives the course of human events is not a titanic struggle between gods and demons, for the Bible with its austere and fierce monotheism dismisses the existence of demons, but rather the recalcitrant nature of man. No need for gargoyles; the demon is within. The field of battle is the human soul, and the contending armies do not hail from foreign lands. Quickly man drives God to distraction and despair, yes, even to repudiate the work of His hands. "How much cosmic agony in this divine attestation of failure!" to quote Professor Yochanan Muffs, who has taught us to read the Bible more as a book about the maturing of God than the education of man.[22] The Hebrew Bible is nothing if not a single unbroken, heroic struggle by God to protect man from himself. Or as the signage above the life-size mirror in the gorilla house of the Bronx Zoo says with a sting when you stand in front of it after having beheld some of the world's most intimidating beasts: "You are looking at the most dangerous animal in the world!" Man himself turns demonic as he rises ever again in rebellion against the will of God. And thus in the spirit of Ecclesiastes the daily liturgy of Judaism intones softly: "Compared to you [God], all the mighty are nothing, the famous nonexistent, the wise lack wisdom, the clever lack reason. For most of their actions are meaningless, the days of their lives emptiness. Human preeminence over beasts is an illusion when all is seen as futility."[23]

In the face of this stark reality, Judaism erected what Freud, with all the uncanny intuition of the genius, described as a system of instinct renunciation. At bottom, its theology and praxis are one: "a subordination of the sense perception to an abstract idea; a triumph of spirituality over the senses."[24] Long before, a cluster of rabbinic sages had argued in a similar vein that the specific details of the commandments were of no matter to a majestic God beyond all human ken; the system as a whole was meant to discipline and purify man.[25] The goal was not to take flight in a miasma of otherworldliness, to obsess about salvation in a world beyond, or even to extirpate the senses in this world, but rather to effect survival in the here and now through a regimen of plain living. As Moses winds down his peroration, he restates the essential message of his entire teaching: "See, I set before you this day life and prosperity, death and adversity. . . . Choose life. . . ."[26] Indeed, Judaism has a conspicuously underdeveloped notion of the afterlife, which its critics often mistook for the absence of a belief in the soul. Instead, it embraced the ultimate goodness of God's creation and evolved a balanced discipline to respect its harmony as we take of its bounty. And, integral to that ethical system of self-restraint is the vision of man as steward and not overlord, for as the Bible so often avers: the land ultimately belongs to its Creator and we mortals are but His tenants.[27]

There is still a third aspect to my argument for Judaism's relevance to the environmental crisis of our day. Were it but a set of lifeless prohibitions, Judaism would have yielded little human satisfaction. To be sure, the enactment of specific forms of self-denial is ideally done with beauty and fervor, and the unintended consequences are often matters of great weight. But how is the void to be filled? To what purpose is the energy and time gained through withdrawal to be put?

The answer relates to the very essence of Judaism. Whatever else Judaism comprises, it is above all a religious culture that places a premium on study. The written word is the distillate of its revelatory experience, the reading of Scripture, the core of its sabbath service, and the unending interpretation of the canon, the basic mode of its religious thought. Jews became masters of deep-reading and their exegetical literature, so vast and diverse, is a tribute to an existential dialogue that never waned. While literacy served as the portal to piety, regular daily study epitomized the supreme religious value. Maimonides' formulation in his twelfth-century code bespoke the normative ideal:

> Every male Jew is obligated to study Torah, whether he be poor or rich, healthy or afflicted, an adolescent or a man of great age, already failing in strength, even a man dependent on public charity or reduced to begging, even a husband and father of children, all are obligated to set aside time for the study of Torah, both morning and evening, as it is said (Joshua 1:8) "And you shall pore over it [the Torah] day and night. . . ." Until when must one study? Until the day he dies. . . .[28]

By way of an aside, it should be noted that for all its loftiness, this spiritual norm did not come to include Jewish women until our own century, though even now some traditionalists continue to reject the enlargement.

But having said that—the goal of daily study is not about what people should do with their lives, but with their spare time. It is not a standard to turn all Jews into rabbis or to entomb them in a world of ancient books. Rather it is a wholesome plea for the cultivation of an inner life that will counter the strains and seductions of prosperity or oppression. Torah is to be internalized through a form of participatory study that ever expands the meaning of sacred texts as it personalizes them. Exegesis is an active exercise, a partnership between the time-bound and the timeless to express individuality through an archaic and sacred medium. For the Jew, the Bible is a notation scale for the composition of religious music of untold intricacy and variation. Daily study hollows out a pocket of inner tranquillity in which refuge is taken from the excesses and cacophonies of an imperfect world. In the words of the psalmist: "Were not your teaching my delight, I would have perished in my affliction."[29] Or as expressed so delicately by the third-century Palestinian rabbi Resh Lakish: "Everyone who occupies himself with the Torah at night, the Holy One, blessed be He, will encircle him with a chord of grace during the day."[30] The highest value is not self-worship as expressed in the ferocious consumerism of American life, but a stubborn quest for balance and equilibrium: a measure of inwardness and self-restraint that lifts one above the grinding materialism of outward concerns.

According to the theosophic wisdom of Lurianic Kabbalah, before God created the world He contracted Himself into a single, small corner of the preexistent cosmos.[31] Whatever the philosophic impulse behind that arresting idea, its psychological thrust is beyond question. To paraphrase Göethe who in *Wilhelm Meister* observed years later in nearly the same language: "It is in his ability to contract that the master first shows himself." To bring forth anything of value requires a decisive act of withdrawal and sustained concentration. So too does the relentless challenge of daily living. In a society that has made of extravagance a commonplace and of distraction a fiendish art, Judaism at its best holds out one model—often dismissed or abused in reality—for reining in the appetites of human consciousness—a strain of asceticism blended with a love of learning. To learn to live with less in this land of milk and honey will demand of all of us, each in his own way, a rapid expansion of our inner resources.

NOTES

1. Lynn White, Jr., "The Historical Roots of Our Ecological Crisis," *Science*, 155 (10 March 1967): 1203–7.
2. Wendell Berry, *What Are People For?* (San Francisco: North Point, 1990), 37.
3. Saul Lieberman, *Hellenism in Jewish Palestine* (New York: The Jewish Theological Seminary of America, 1950), 83.
4. Gen. 9:2–6; cf. Gen. 1:29–30.
5. Lev. 11; Deut. 14.
6. Yehezkel Kaufmann, *The Religion of Israel*, trans. and abridged by Moshe Greenberg (Chicago: University of Chicago Press, 1960), 180–82.
7. Deut. 25:4, 22:10; Lev. 22:28.
8. Deut. 22:6–7.
9. Isaac Klein, *A Guide to Jewish Religious Practice* (New York: Jewish Theological Seminary of America, distributed by Ktav Pub. House, 1979), 360ff.
10. Exod. 23:5.
11. Isa. 11:1–9; also Hos.2:20.
12. Lev. 25:1–7.
13. Lev. 19:9–10.
14. Lev. 19:19; Deut. 22:10.
15. Num. 18:12–18.
16. Lev. 19:23–25.
17. Deut. 20:19; Maimonides, *Mishne Torah*, Hilkhot Melakhim 6:8–10.
18. *Mishna*, Yevamot 6:6.
19. Klein, *Guide to Jewish Religious Practice*.
20. Quoted by Pico in his "Oration on the Dignity of Man," in Ernst Cassirer et al., eds., *The Renaissance Philosophy of Man* (Chicago: University of Chicago Press, 1948), 223.
21. Quotation from Ps. 8:6.
22. Yochanan Muffs, "Of Image and Imagination in the Bible," in *J. James Tissot, Biblical Paintings* (New York: Jewish Museum, 1982), 9.
23. Jules Harlow, ed., *Siddur Sim Shalom* (New York: Rabbinical Assembly: United Synagogue of America, 1985), 13.
24. Sigmund Freud, *Moses and Monotheism* (New York: Vintage Books, 1959), 144.
25. C. G. Montefiore and H. Loewe, *A Rabbinic Anthology*, rpt. 3d. rev. ed. (New York: Schocken Books, 1974), 148–49.

26. Deut. 30:15, 19.
27. Lev. 25:23.
28. Maimonides, *Mishne Torah*, Hilkhot Talmud Torah, 1:8, 10.
29. Ps. 119:91.
30. Babylonian Talmud, Hagiga, 12b.
31. Gershom G. Sholem, *Major Trends in Jewish Mysticism*, 3d rev. ed. (New York: Shocken Books 1954), 260–64.

CHAPTER 3

A Square in the Quilt: One Theologian's Contribution to the Planetary Agenda

Sallie McFague

Fig. 3. Georgia O'Keeffe (1887–1986) evokes in her images of plants and flowers the same dimensions that make her paintings of landforms so arresting. Leaves of a Plant *(1943) at once registers vast and unceasing currents of energy, conveys the spiritual significance of a particular physical form, and offers a vision of sensuality that is electrically present for us in our humanity. (Collection of Mr. and Mrs. Gerald Peters, Santa Fe, New Mexico.)*

Throughout the nineteenth and twentieth centuries liberal Christian theologians have been involved in reconstructing the idea of God in response to new intellectual and social developments. At the end of the twentieth century, this process of adjustment continues with fresh energy due to the influence of global movements of democratic social liberation, new ecological and holistic modes of scientific understanding, interfaith and cross-cultural dialogue, and a realization that humanity today has the power to destroy all life on earth. These realities form the context in which Professor Sallie McFague does Christian theology.

Sallie McFague's work also reflects the convergence today of ecological and feminist modes of thought. Feminist thinkers have become especially sensitive to issues of environmental degradation, perceiving a parallel between the conquest and exploitation of nature, which has historically been envisioned as female, and the subjugation and abuse of women. Feminist theology, which has generated some of the most searching and creative theological reflection in recent decades, is centrally concerned with criticizing and replacing images of God that foster attitudes of conquest and domination. Sharing this objective, McFague seeks to free the powers of imaginative religious vision and to construct new liberating images of God for our time.

In her essay, Sallie McFague presents a Christian theologian's vision of interdependence. Drawing on the insights of cosmology, ecology, and feminism, she offers a critique of traditional monarchical and patriarchal images of God and seeks to reconstruct the way Christians understand the interrelation of God, humanity, and nature. She places special emphasis on the spiritual significance of "the common creation story" emerging from contemporary science. Endeavoring to remain faithful to the heart of the Christian Gospel, her theology shifts the emphasis from dualism to holism and from authority and hierarchy to participation and community. Central to McFague's purpose is a concern to "help people to think and to act holistically" as "citizens of the planet."

IN a recent book on the economy and the environment, the authors close their four-hundred-page study of how we must change our systems of production and consumption on a grim note: they note how late in the day it is, how bitter one becomes thinking of missed opportunities, and how hard it is to avoid resentment toward those who block needed change. They conclude with these words:

> Yet there is hope. On a hotter planet, with lost deltas and shrunken coastlines, under a more dangerous sun, with less arable land, more people, fewer species of living things, a legacy of poisonous wastes, and much beauty irrevocably lost, there is still the possibility that our children's children will learn at last to live as a community among communities. Perhaps they will learn also to forgive this generation its blind commitment to even greater consumption. Perhaps they will even appreciate its belated efforts to leave them a planet still capable of supporting life in community.[1]

Seldom have we been summoned to a more discouraging task. We are not being told that we can make things better or even that we can hold onto what we have; rather, we are being warned that unless we change our thinking and our actions immediately and radically we may not even have a planet suitable for decent living, that is, "life in community." This is indeed a sobering thought, for it appears not to be enough to celebrate the intricate dance of life in all its beauty and richness that we have belatedly come to appreciate and know that we belong to. The real sting comes when one takes to heart the revolution this involves in one's daily life, at least for first-world people. It is the particular, concrete daily restrictions and diminishments that cause most people to rationalize and temporize on ecological matters.

But the issue is deeper still. We do not want to admit how dire things are or how much we are to blame. And yet, if the environmental crisis is to be anything more than another fad, which it has the potential of becoming as oil companies, plastic manufacturers, and car makers run ads boasting of their ecological sensitivity, we must both acknowledge the seriousness of the situation and our responsibility for it. In ways that have never before been so clear and stark, we have met the enemy and know it is ourselves. While the holistic, planetary perspective leads some to insist that all would be well if only a "creation spirituality" were to replace the traditional "redemption spirituality" of the Jewish and Christian traditions, the issue, I believe, is not that simple. It is surely the case that an overemphasis on redemption to the neglect of creation needs to be redressed; moreover, there is much in the common creation story that calls us to a profound appreciation of the wonders of our being as well as the being of all other creatures. Nonetheless, it is doubtful if that knowledge and appreciation will be sufficient to deal with the exigencies of our situation.

The enemy, identified as indifferent, selfish, short-sighted, xenophobic, anthropocentric, greedy human beings, calls, at the outset, for a renewed emphasis on sin as the cause of much of the planet's woes and, therefore also, for a broad and profound repentance. Sin is understood in the Christian tradition, as Augustine said centuries ago, to be "living a lie," living in false relations to God and other beings. It is, as he said in a term that may sound quaint and anachronistic but which is ecologically up-to-date, "concupiscence," an insatiable appetite which causes one to want to "have it all" for oneself. Sin is assuming one is the center and that all others exist for one's benefit. Sin is being out of proportion, out of relation, in terms of the ways things really are; sin is "living a lie." Sin, therefore, is hardly an old-fashioned or dated concept as some would have us suppose; rather, it is the first step toward health and reality, toward living rightly, living in proper, appropriate relations with all other beings. It is the first step in which we acknowledge the myriad ways that we personally and corporately have ruined and continue to ruin God's splendid creation—acts which we and no other creature can knowingly commit. Human responsibility for the fate of the earth is a recent and terrible knowledge; our loss of innocence is total, for we *know* what we have done.

A profound acknowledgment of our complicity in the deterioration of our planet is, I believe, the first step. This leads to a second step: the sense of responsibility for helping to preserve a planet still capable of supporting life in community. While the call to repentance may be a peculiarly theological or religious contribution to our planetary crisis, the challenge to contribute constructively to the earth's well-being is a universal task, that is, a task for everyone. We, all of us, are being called to do something unprecedented. We are being called to think about "everything that is," for we now know that everything is interrelated and that the well-being of each is connected to the well-being of the whole. This suggests a "planetary agenda" for all the religions, all the various fields of expertise, all those who help the poor and the outcast. This is not a vocation for theologians or any other group of professionals alone, although they, along with all others, are called to this task. It is the universal vocation for all people. *The* moral issue of our day is the global one of whether we and other species will live and how well we will live. Therefore, the moral challenge, the planetary vocation or agenda, can be summarized by the rallying cry of the World Council of Churches: peace, justice, and the integrity of creation. These ideals, which at one time might have appeared separate or at odds with each other, must now be seen as profoundly interrelated if *any* of them is to be achieved.

Ecological and justice issues have sometimes been viewed as in competition for attention, or scarce goods, or both. Increasingly, however, I believe that perception is short-sighted. Liberation theologies have been concerned with how we can change the world; ecological theologies are concerned with how we can save the world. This may seem like a major difference, but I think it is not. Rather, the focus of the liberation theologies has widened to include, in addition to all oppressed human beings, all oppressed creatures as well as the ecosystem that supports all forms of life.

Liberation theologies insist, rightly I believe, that all theologies are written from particular contexts. The one context which has been neglected and is only now emerging is the broadest and most basic one. It is the context of the planet, a context which we all share and without which we cannot survive. Hence, it seems to me that this latest shift in twentieth-century theology is not away from the concerns of the liberation theologies, but toward a deepening of those concerns, toward a recognition that the fate of the oppressed and the fate of the earth are inextricably interrelated, for we live, all of us, on one planet, a planet we increasingly realize is fragile and vulnerable to our destructive behavior.

The interlinking of justice and ecological issues becomes evident when one considers the characteristic dualistic, hierarchical mode of Western thought in which a superior and an inferior are correlated: male/female, white/people of color, heterosexual/homosexual, able-bodied/physically challenged, culture/nature, mind/body, human/nonhuman. These correlated terms, which are seen not as merely different but as normatively ranked, reveal clearly that domination and destruction of the natural world are inexorably linked with the domination and oppression of the poor, people of color, and all others that fall onto the "inferior" side of the correlation. Nowhere is this more clear than in the ancient and deep identification of women with nature, an identification so profound that it touches the very marrow of our being: our birth from the bodies of our mothers and our nourishment from the body of the earth. The power of nature—and of women—to give and withhold life epitomizes the inescapable interconnection of the two and thus the necessary relationship of justice and ecological issues. As many have noted, the status of women and of nature has been historically parallel: as one goes, so goes the other.

A similar correlation can be seen between other forms of human oppression and a concern for the natural world. The most obvious connection, of course, is that unless ecological health is maintained, the poor and others with limited access to scarce goods due to race, class, gender, or physical capability, cannot be fed. Grain must be grown for all to have bread. That concern, of course, is central, but it is also the case that the characteristic Western mind-set that has accorded intrinsic value, and hence duties of justice, principally to the upper half of each dualism, has considered it appropriate and proper to use those on the lower half for its own benefit. For instance, Western multinational corporations regard it as "reasonable" and "normal" to use third-world people and their natural resources for their own financial benefit, at whatever cost to the indigenous peoples and the health of their lands.

Thus, the planetary agenda to think holistically, to think of "everything that is," is neither exaggerated nor romantic. It is, on the contrary, necessary and realistic in the sense that it is what we must do, given the interconnections and interdependencies that we now understand to be the nature of reality. Needless to say, this is indeed an agenda for *everyone*, as each of us finds a way to act locally as we think globally, finds the issue or project or concern to which we will devote our time and talents, or as feminists like to put it, sews the square which each contributes to the common quilt. Collegiality *and* difference, the acceptance of a planetary agenda with the recognition that different voices, different tasks, different understandings are constitutive of that agenda, is the modality appropriate to our time. If theologians, for example, were to accept this context and agenda for their work, they would see themselves in dialogue with all those in other areas and fields similarly engaged: those who feed the homeless and fight for animal rights; the cosmologists who tell us of the common origins (and hence interrelatedness) of all forms of matter and life; economists who predict how we must change if the earth is to support its population; the legislators and judges who work to advance civil rights for those discriminated against in our society; the Greenham women who picket nuclear plants and the women of northern India who literally "hug" trees to protect them from destruction and so on and on.

In summary, theology becomes an "earthly" affair in the best sense of the word; that is, it becomes an aid in helping people to live rightly, appropriately, on the earth, in our home. It is, as the Jewish and Christian traditions have always insisted, concerned with "right relations," relations with God, neighbor, and self, but now the context has broadened to include what has sometimes dropped out of the picture, especially in the last few hundred years—the oppressed neighbor, the other creatures, and the earth that supports us all. This shift could be seen as a return to the roots of the Judeo-Christian tradition, a tradition that has insisted on the creator and redeemer God as the source and salvation of *all* that is. We now know that "all that is" is vaster, more complex, more awesome, more interdependent than any other people has ever known. The new theologies that emerge from such a context have the opportunity to view divine transcendence in more awesome as well as in more intimate ways than ever before. They also have the opportunity, indeed, the obligation, to understand the place of human beings on our planet as one of radical interrelationship and interdependence with all other forms of life, as well as special responsibility for their well-being.

I would like to become very specific. What is my little "square" that I offer to the common quilt? What can I *as a theologian* in the Christian tradition do constructively so that our planet can continue to support life in community? I emphasize "as a theologian," because I believe that the planetary agenda cannot be an avocation, something one does in addition to one's everyday work—a pastime or hobby, as it were—but needs to be one's *vocation*, one's central calling. It is perhaps obvious how raising children, gardening, teaching, nursing, or caring for animals might contribute to the planet's agenda, but how does theology (let alone business, law, housekeeping, plumbing, or car manufacturing)? I leave it to each of those professions to imagine how they *might* fit, for I believe they in fact *must* fit (if not, their legitimacy is in question). But those who practice these arts and skills ought to be the ones who say in what ways they do fit or ought to change in order to do so.

There are many different theological tasks relevant to planetary well-being. One of central importance is learning to think differently about ourselves, others, and our planet because learning to think differently usually precedes being willing and able to act differently. Much of one's thinking at the basic level of worldview and one's place in it is derived from the dominant images in the religious traditions of one's culture. This "thinking" is not primarily conscious nor is it limited to active members of a religious tradition. Western culture was and still is profoundly formed by the Hebrew and Christian religions and their stories, images, and concepts regarding the place of human beings, history, and nature in the scheme of things. Moreover, I believe it is the major images or metaphors of a tradition that influence behavior more powerfully than its central concepts or ideas. For instance, it is the image of God as king and lord rather than the idea of God as transcendent that has entered most deeply into Western consciousness.

Let me briefly suggest some ways in which the modern Western worldview has been deeply influenced by Christianity, and especially by Protestant Christianity. As many have pointed out, the Christian tradition is and has been not only deeply androcentric (centered on males and male imagery for the divine) but also deeply anthropocentric (focused on human well-being) to the almost total neglect of other species and the natural world, especially during the last few hundred years. It is also focused on redemption narrowly conceived, on human salvation, often understood in individualistic and otherworldly terms. The creation and health of *all*, of the earth and its creatures, has seldom been a central concern of the tradition. Moreover, its dominant imagery has been monarchical. God is imaged as king, lord, and patriarch of a kingdom which he rules, a kingdom hierarchically ordered. God has all the power in this picture, with human beings seen as rebellious, prideful sinners against the divine right.

Needless to say, this picture is not what thoughtful Christians or other thoughtful people influenced by Christian culture hold consciously, but its main tenets have seeped into the Western worldview to the extent that most Westerners, quite unselfconsciously, believe in the sacredness of every individual human being (while scarcely protesting the extinction of all the members of other species); believe males to be "naturally" superior to females; find human fulfillment (however one defines it) more important than the well-being of the planet; and picture God (whether or not one is a believer) as a distant, almighty superperson. This dualistic, hierarchical picture supports another form of dangerous behavior: the superiority of one's own nation over others and hence the validation of a nationalistic, militaristic, xenophobic horizon. Christianity is surely not alone responsible for this worldview, but to the extent that it has contributed to and supported it, the deconstruction of some of its major metaphors and the construction of others is in order.

The portrayal of God as monarch ruling over his kingdom is the dominant model in Jewish, Catholic, and Protestant thinking and is so widely accepted that it often is not recognized as a picture, that is, a construction of the divine/human relationship. To many, God is the lord and king of the universe.[2] Yet the monarchical model has been thoroughly and roundly criticized not only by feminists but by a host of other theologians as well. It is not necessary to review the criticism of the model here, except for a few points. In the monarchical model, God is distant from the world, relates only to the human world, and controls that world through domination and benevolence.[3]

On the first point: the relationship of a king to his subjects is necessarily a distant one, for royalty is "untouchable." God as king is in his kingdom, which is not of this earth, and we remain in another place, far from his dwelling. In this picture, God is worldless and the world is Godless—the world is empty of God's presence. Whatever one does for the world is finally not important in this model, for its ruler does not inhabit it as his primary residence and his subjects are well advised not to become too enamored of it either. At the most, the king is benevolent, but this benevolence extends only to human subjects.

And this is the second point: as a political model focused on governing human beings, it leaves out the entire rest of the earth and its many creatures. It is simply blank as to the natural world and hence has encouraged a similar indifference in human beings. God's kingdom is composed exclusively of human beings.

Finally, in this model, God rules either through domination or benevolence, thus undercutting human responsibility for the earth. It is simplistic to blame the Jewish and Christian traditions for the ecological crisis, as some have done, on the grounds that Genesis instructs human beings to have "dominion" over nature; nevertheless, the imagery of sovereignty supports attitudes of control and use toward the nonhuman world. Whatever might have been nature's superiority in the past, the balance of power has shifted to us. Extinction of species by nature, for instance, is in a different dimension from extinction by design, which only we can bring about. The model is lacking even if God's power is seen as benevolent rather than as domineering. Then it is assumed that all will be well, that God will care for the world with no help from us. The heavenly father will take care of his children; we can leave it up to him.

The images in this model are constructions and as such they are partial, relative, and inadequate. They are metaphors abstracted from human relations (relations with kings, lords, masters) and applied to God. Hence, they are in no sense "descriptions" of God; yet their power is deep and old, their influence inscribed into our being from our earliest years. They are, therefore, difficult to discard. Yet, as we have seen, the monarchical model is dangerous in our time, for it encourages a sense of distance from the world, is concerned only with human beings, and supports attitudes of either domination of the world or passivity toward it. This chilling realization adds a new importance to the images we use to characterize our relationship to God, to others, and to the nonhuman world. No matter how ancient a metaphorical tradition may be, and regardless of its credentials in scripture, liturgy, and creedal statements, it still must be discarded if it threatens the continuation and fulfillment of life. If, as I believe, the heart of the Christian gospel is the salvific power of God for all of creation, triumphalistic metaphors cannot express that reality *in our time* whatever their appropriateness may have been in the past.

What are other possibilities for imaging God's relationship to the world and our place within it? The first question one must ask is: what world? Probably the single most important thing that theologians can do for the planetary agenda is to insist that the "world" in question, the world in which to understand both God and human beings, is the contemporary scientific picture of the earth, its history, and our place in it that is emerging from cosmology, astrophysics, and biology. Neither the world of the Bible, nor of Newtonian dualistic mechanism, nor of present-day creationism is the world to which we must respond as theologians. I am not suggesting in any sense that science dictate to theology nor that the two fields be integrated. I am making a much more modest, though critically important, proposal. Contemporary theology, if it is to help people to think and act holistically, must make its understanding of the God/world relationship consonant with contemporary views of reality. A theology that is not commensurate with reality as culturally understood is not credible. Moreover, the contemporary view coming from the sciences is so awesome, rich, and provocative for imaging both divine and human relationships that the political models seem pale and narrow in comparison.

In broad strokes, the common creation story emerging from the various sciences claims that some fifteen or so billion years ago the universe began with a big bang, exploding matter, which was intensely hot and infinitely concentrated, outward to create some hundred billion galaxies of which our galaxy, the Milky Way, is one, housing our sun and its planets. From this beginning came all that followed so that everything that is is related, woven into a seamless network, with life gradually emerging after billions of years on this planet (and probably on others as well), and evolving into the incredibly complex and beautiful earth that is our home. All things living and all things not living are the products of the same primal explosion and evolutionary history, and hence are interrelated in an internal way from the very beginning. We are distant cousins to the stars and near relations to the oceans, plants, and all other living creatures on our planet.

The characteristics of this picture suggest possibilities for understanding both God and human existence that are radically different from what we found in the monarch/realm worldview. The "world" here is, first of all, the universe, beside which the traditional range of divine concern with, mainly, human subjects dwindles, to put it mildly. In this view, God relates to the entire fifteen-billion year history of the universe and all its entities, living and nonliving. On the "clock" of the universe, human existence appears a *few seconds* before midnight. This suggests, surely, that the whole show could scarcely have been put on for our benefit; our natural anthropocentrism is indeed sobered. Nevertheless, it took fifteen billion years to evolve creatures as complex as human beings; hence, the question arises of our peculiar role in this story, especially in relation to our own planet.

A second feature of the common creation story is the radical interrelatedness and interdependence of all aspects of it, a feature of utmost importance to the development of an ecological sensibility. It is *one* story, a *common* story, so that everything that is traces its ancestral roots within it, and the closer in time and space entities are, the closer they are related. Thus, while we may rightly feel some distance from such "relatives" as exploding stars, we are "kissing cousins" with everything on Planet Earth and literally brothers and sisters to all other human beings. Such intimacy does not, however, undercut difference; in fact, one of the outstanding features of postmodern science's view of reality is that individuality *and* interdependence characterize everything. It is not just human beings that have individuality, for the veins on every maple leaf, the configuration of every sunset, and the composition of every pile of dirt is different from every other one. This portrayal of reality undercuts notions of human existence as separate from the natural, physical world; or of human individuality as the only form of individuality; or of human beings existing apart from radical interdependence and interrelatedness with others of our own species, with other species, and with the ecosystem. The continuity of nonliving and living matter displays another crucial feature: the inverse dependency of the most complex entities on the less complex. Thus, the plants can do very nicely without us, but we would perish quickly without them. The higher and more complex the level, the more vulnerable it is and the more dependent upon the levels that support it. Again, we see implications for reconceiving the "place" of human beings in the scheme of things.

Another feature of the common creation story is its public character; it is available to all who wish to learn about it. Other creation stories, the cosmogonies of the various world religions, are sectarian, limited to the adherents of different religions. But any person on the planet has potential access to the common creation story and simply as a human being is included in it. This common story is available to be remythologized by any and every religious tradition and hence is a place of meeting for the religions, whose conflicts in the past and present have often been the cause of immense suffering and bloodshed. What this common story suggests is that our primary loyalty should not be to nation or religion, but to the earth and its Creator (albeit that Creator may be understood in different ways). We are members of the universe and citizens of Planet Earth. Were that reality to sink into human consciousness all over the world, not only war among human beings but ecological destruction would have little support in reality. This is not to say they would disappear, but those who continued in such practices would be living a lie, that is, living in a way that is out of keeping with reality as it is currently understood.

Finally, the common creation story is a *story*: it is a historical narrative with a beginning, middle, and presumed end, unlike the Newtonian universe which was static and deterministic. It is not a "realm" belonging to a king, but a changing, living evolving event (with billions of smaller events making up its history). In our new cosmic story, time is irreversible, genuine novelty results from the interplay of chance and necessity, and the future is open. This is an unfinished universe, a dynamic universe, still in process. Other cosmologies, including mythic ones such as Genesis and even the earlier scientific ones, have not been historical, for in them creation was "finished." Rather than viewing God as an external, separate being ruling over the world, it is appropriate to see God as in, with, and under the entire evolutionary process. Paul's statement that God is the one "in whom we live and move and have our being" takes on new and profound significance. In this picture God would be understood as a *continuing* creator, but of equal importance, we human beings might be seen as co-creators, as the self-conscious, reflexive part of the creation that can participate in furthering the process.

To summarize: the characteristics of the common creation story suggest a decentering and recentering of human beings. We are radically interrelated with and dependent on everything else in the universe and especially on our planet. We exist as individuals in a vast community of individuals within the ecosystem, each of which is related in intricate ways to all others in the community of life. We exist with all other human beings from other nations and religions within a common creation story that each of us can know about and identify with. The creation of which we are a part is an ongoing, dynamic story which we alone (so we believe) understand, and hence we have the potential to help it continue and thrive, or to let it deteriorate through our destructive, greedy ways.

Our position in this story is radically different from our place in the king/realm picture. We are decentered as the only subjects of the king and recentered as those responsible for both knowing the common creation story and being able to help it flourish. In this story we feel profoundly connected with all other forms of life, not in a romantic way, but in a realistic way. We *are* so connected and hence we had better *live* as if we were. We feel deeply related, especially, to all other human beings, our closest relatives, and realize that *together* we need to learn to live responsibly and appropriately in our common home.

If this kind of thinking became widespread—thinking of ourselves as citizens of the planet, breaking down all forms of parochialism—some of the things that must happen if we and the earth are to survive and flourish, might be able to. Once the scales have fallen from one's eyes, once one has seen and believed that reality is put together in such a fashion that one is profoundly united to and interdependent with all other beings, everything is changed. One sees the world differently: not anthropocentrically, not in a utilitarian way, not in terms of dualistic hierarchies, not in parochial terms. One has a sense of belonging to the earth, having a place in it along with all other creatures, and loving it more than one ever thought possible.

Such a perspective does not diminish either human beings or God; in fact, both are enlarged. Human beings have been decentered as the point and goal of creation, and recentered as co-creators of it; God has been decentered as king of human beings and recentered as the source, power, and goal of the fifteen-billion-year history of the universe. As ethicist James M. Gustafson puts it, while we are not the "measure" of creation, we are its "measurer."[4] We are not the center or the point of creation, but we are the only ones, to our knowledge, who know the story of creation. In fact, we human beings presently alive are the first human beings who really know this story because only within the last fifty years has it gradually emerged from the scientific community. We alone know this awesome fact, and the more one knows about this story—the micro and macro worlds that surround our middle world, the worlds of the very tiny and the unimaginably immense and ancient—the more filled with wonder one becomes. This is not to suggest that an aesthetic response is the principal one. Wonder and awe at the immensity and age of the universe can generate a sense of diffidence toward puny earthly problems. A genuine aesthetic response is necessarily accompanied by an ethical one; that is, our responsibility for preserving the beauty, diversity, and well-being of our tiny corner of the universe, Planet Earth. As those responsible for helping the creative process to continue and thrive on our planet, we can scarcely imagine a higher calling. We have been recentered as co-creators.

In this picture of God and the world, *God* is certainly not diminished. To think of God as creator and continuing creator of this massive, breathtaking cosmic history, makes all other traditional images of divine transcendence, whether political or metaphysical, seem small indeed. The model of God as king is, by comparison, "domesticated transcendence," for a king rules only over human beings. A genuinely transcendent model would insist that God is the source, power, and goal of the total universe, but a source, power, and goal that works within its natural processes; hence, the model, while genuinely transcendent, is also profoundly immanent. The king/realm model is neither genuinely transcendent (God is king over one species recently arrived on a minor planet in an ordinary galaxy) nor genuinely immanent (God as king is an external superperson, not the source, power, and goal of the entire universe).

The common creation story is a rich resource for reimaging the God of the Christian tradition, who is understood to be both radically transcendent and radically immanent: both the one and only God as well as the one "in whom we live and move and have our being." This God is the creator and savior of all that exists, the one who gives life and restores health to stricken and wounded life. The psalms of the Hebrew scriptures often express these qualities magnificently, as in these lines from Psalm 104:

> O Lord, how manifold are your works!
> In wisdom you have made them all;
> the earth is full of your creatures.
> Yonder is the sea, great and wide,
> creeping things innumerable are there,
> living things both small and great.
> . . . .
> These all look to you
> to give them their food in due season;
> when you give to them, they gather it up;
> when you open your hand, they are filled with good things.

The predominance of the king/realm imagery during the last several hundred years, with its focus on human salvation, is not representative of the entire Jewish and Christian traditions. This tradition has a wider, more comprehensive dimension that has at times seen human well-being in the context of the well-being of the entire creation. With the common creation story we now have a resource for reimaging and radicalizing God as creator and redeemer of all that is, with human well-being as one important though by no means the only focus of divine concern.

Within this story we human beings find, once again, our proper place within the whole, a place that, with the rise of modern science and its wedding to technology, seemingly giving us control over the natural world, we have forgotten. Yet now our place involves more responsibility than ever before, because we know that we have power, not the power to create but the power to destroy. We realize that it is only by living appropriately, in proper relations with all other beings, that we can fulfill our responsibilities to the well-being of creation. Within this story, God is, once again, the source, power, and goal of all that is, the creator and redeemer of the cosmos, and not merely king of human beings. Yet now God is giver and renewer of a universe so vast, so old, so diverse, so complex that all earlier and other images of divine glory and transcendence are dwarfed by comparison. And yet, this transcendence is one immanental to the universe, for God is not a superperson, a king, external to cosmic processes, but the source, power, and goal immanent in these processes.

One of the rallying cries of the Protestant Reformation was "let God be God." That is precisely what the common creation story, as a resource for theology, suggests. It is God and not we who creates and makes whole. We, at the most, are helpers, whose proper place within the whole on which we depend is to acknowledge who we are in reality and where we fit. The common creation story tells us that the earth is our home; we belong here, and we have responsibilities to our home.

It is precisely this sense of belonging, of being at home, that is the heart of the matter. It is the heart of the matter because it is the case: we do belong. As philosopher Mary Midgley writes, "We are not tourists here. . . . We are at home in this world because we were made for it. We have developed here, on this planet, and we are adapted to life here. . . . We are not fit to live anywhere else."[5] Postmodern science allows us to regain what late medieval culture lost at the Reformation and during the rise of dualistic mechanism in the seventeenth century: a sense of the whole and where we fit in it. Medieval culture was organic, at least to the extent that it saw human beings, while still central, as embedded in nature and dependent upon God. For the last several centuries, for a variety of complex reasons, we have lost that sense of belonging. Protestant focus on the individual and otherworldly salvation, as well as Cartesian dualism of mind and body, divided what we are now trying to bring back together, and what must be reintegrated if we and other beings are to survive and prosper. But now, once again, we know that we belong to the earth, and we know it more deeply and thoroughly than any other human beings have ever known it. The common creation story is more than a scientific affair; it is implicitly, deeply moral, for it raises the question of the place of human beings in nature, and calls for a kind of praxis in which we see ourselves in proportion, in harmony, and in a fitting manner relating to all others that live and all systems that support life.

To *feel* that we belong to the earth and to accept our proper place within it is the beginning of a natural piety, what Jonathan Edwards called "consent to being," consent to what is. It is the sense that we and all others belong together in a cosmos, related in an orderly fashion, one to the other. It is the sense that each and every being is valuable in and for itself, and that the whole forms a unity in which each being, including oneself, has a place. It involves an ethical response, for the sense of belonging, of being at home, only comes when we accept our proper place and live in a fitting, appropriate way with all other beings. It is, finally, a religious sense, a response of wonder at and appreciation for the unbelievably vast, old, rich, diverse, and surprising cosmos, of which one's self is an infinitesimal but conscious part, the part able to sing its praises.

To summarize: one square in the quilt, one contribution to the planetary agenda, is the deconstruction of models and metaphors that are oppressive and dangerous to our planet as well as the suggestion of alternatives. Since the Christian tradition has contributed a number of problematic images, it is incumbent upon its theologians to analyze and criticize such models and metaphors. It is also the responsibility of theologians to suggest, from current resources, alternative models and metaphors to express the relation of God to the cosmos as well as the place of human beings in the cosmos. The common creation story is one such rich resource for reimaging both divine and human reality in relation to the universe, and especially to our planet.

The planetary agenda is the most serious, most awesome fact facing us; it is concerned with whether we live or die and how well we live, if we do live—questions usually reserved for religions and their solutions to issues of mortality and salvation. We see now, however, that health and well-being are profoundly "earthly" affairs—while still being religious ones, with all the urgency of religious questions. It is no exaggeration to say that the planetary agenda is a life and death matter. We now know this and we know that our time is limited to do what needs to be done. The planetary agenda is everyone's agenda. Each of us is called upon to contribute one square to the quilt. As time is short, we had better get about the business of doing so.

NOTES

1. Herman E. Daly and John B. Cobb, Jr., *For the Common Good: Redirecting the Economy toward Community, the Environment, and a Sustainable Future* (Boston: Beacon Press, 1989), 400.
2. "The *monarchical model* of God as King was developed systematically, both in Jewish thought (God as Lord and King of the Universe), in medieval Christian thought (with its emphasis on divine omnipotence), and in the Reformation (especially in Calvin's insistence on God's sovereignty). In the portrayal of God's relation to the world, the dominant western historical model has been that of the absolute monarch ruling over his kingdom" (Ian G. Barbour, *Myths, Models and Paradigms: A Comparative Study in Science and Religion* [New York: Harper and Row, 1974], 156).
3. For a more complete critique of the monarchical model, see my book, *Models of God: Theology for an Ecological, Nuclear Age* (Philadelphia: Fortress Press, 1987), chap. 3.
4. James M. Gustafson, *Ethics from a Theocentric Perspective, Volume I: Theology and Ethics* (Chicago: University of Chicago Press, 1981), 28.
5. Mary Midgley, *Beast and Man: The Roots of Human Nature* (Ithaca, N.Y.: Cornell University Press, 1978), 194–95.

CHAPTER 4

Liberal Democracy and the Fate of the Earth

J. Ronald Engel

Fig. 4. James Nachtwey's (1948–) photograph, The Birdman of Harlem *(1989), portrays Craig, one of a small group found on New York City roof tops who raise and train flocks of pigeons in acrobatic flying. Caught in an oppressive urban industrial network, the whole earth reaches with Craig for the sky, the light, the freedom of birds—the freedom to be. (Magnum Photographs, New York.)*

At the heart of the American democratic tradition lies a set of ethical and social values that center around the principles of freedom, equality, and community. As American democratic life has developed, it has steadily expanded its sense of moral community granting freedom to enslaved African-Americans and rights to women and others who were originally excluded from full participation. For some people like Walt Whitman, Jane Addams, and John Dewey, the tradition of moral democracy has been a matter of religious faith and commitment. They viewed the democratic way of life as a vehicle of personal growth and social transformation, leading to a uniting of the divine and the human. In this outlook, democracy is most fundamentally an individual way of living in which the gulf between the sacred and the secular is overcome.

In addition, there has been a tradition of holistic and ecological thinkers, including the literary naturalist Henry David Thoreau, the animal rights activist Henry S. Salt, and more recently Gary Snyder, Lynn White, Jr., and Rosemary R. Ruether, who have argued that our sense of democratic community should be expanded to include the whole community of life. These radical democrats would extend the concept of natural rights to include the rights of nature, an outlook consistent with that of many Native American traditions.

J. Ronald Engel has deep roots in the American tradition of spiritual democracy, and he works on a frontier today where the values of democracy, ecology, and religious faith converge. He has also been a leader in the international deliberations that have focused on the concept of sustainable development. His contributions have focused on creation of a global ethic of sustainability that integrates the values of social and economic justice with the values of justice for nature. "Nature and humanity will be liberated together or not at all," he asserts.

The title of Professor Engel's essay reflects his conviction that the environmental crisis is a crisis of citizenship and can be addressed only by a renewal of the liberal democratic faith, which he interprets as "comprehensive personal, social, and religious faith." Fundamental to the concepts of ecology, citizenship, and liberal democracy, argues Engel, is the idea of the individual in community. Democratic citizenship in an ecological age means responsibility to and for the community in its most universal form. To awaken to the liberal democratic faith, which identifies the world of everyday relations as "the sphere of ultimacy," is "to awaken to Spirit in Nature."

FACING the title page of *Walden*, Henry David Thoreau placed these now well-known words: "I do not propose to write an ode to dejection, but to brag as lustily as chanticleer in the morning, standing on his roost, if only to wake my neighbors up."

Awakening to Spirit in Nature was a basic theme for Thoreau. In his essay "Walking," he writes:

> Unless our philosophy hears the cock crow in every barn-yard within our horizon, it is belated. That sound commonly reminds us that we are growing rusty and antique in our employments and habits of thought. . . . It is an expression of the health and soundness of Nature, a brag for all the world. . . . Where he lives no fugitive slave laws are passed. Who has not betrayed his master many times since last he heard that note?[1]

It is no less a basic theme in the life of a contemporary writer and conscience of his people, Vaclav Havel, president of Czechoslovakia. As he remarked in his address to the Polish Congress in February of 1990: "We have awakened; now we must awaken those in the West who have slept through our awakening."[2] It is said Havel wears a parka with the word "Awakenings" across the back.

From time to time voices are raised with the power to awaken us to our proper place and responsibility in the order of things. Just as the history of social and environmental protest in the past century cannot be understood apart from the author of *Walden* and "Civil Disobedience," the political history of the next century will not be understood apart from the playwright-author of *Temptation* and *The Power of the Powerless*.

There are differences between these two men. Thoreau spent one night in jail for his conscience, Havel almost five years in prison. Thoreau eschewed active political life; Havel shouldered the public role destiny chose for him. Thoreau's heritage was the English Revolution of 1649, the American Revolution of 1776, and the flowering of New England in the 1830s. Havel's heritage is the movement for religious tolerance and national renewal led by the Czechoslovakian martyr Jan Hus in the fourteenth century, the humanism of Tomáš Masaryk, founder of the Czechoslovakian republic in 1918, and the famous Prague Spring of 1968.

But our task is to discern broad redemptive movements at work in the world today, and from this perspective the common political and religious heritage these two men share is far more significant than their differences.

Thoreau and Havel stand in the liberal democratic tradition. But unlike many persons who might claim that identity, they believe that the personal, social, political, and economic principles of liberal democracy are more than matters of expedience; rather, they express important moral and spiritual truths about the ultimate meaning of life. They hold in common the "liberal democratic faith."

Those of us who share this faith, and try to talk about it, have difficulty doing so. We all know, or think we know, what Christianity, or Islam, or Judaism, *is*—but liberal democracy as a faith or religion? Isn't democracy simply how we elect our government? We are in the same situation as Thoreau when he said that he had walked into many lyceums, and "done my best to make a clean breast of what religion I have experienced, and the audience never suspected what I was about."[3]

The challenge is especially great when one believes, as I do, that the liberal democratic faith is not something some people have and others do not, but that we all have some experience of it, and a stake in how successfully it is articulated.

The challenge is even greater when one is not only concerned to sing the praises of liberal democracy, but to criticize and reform it, as I am. With Thoreau and Havel, I have grave reservations about liberal democracy as it is known and practiced in contemporary life, especially the life of our modern industrial and consumer way of life masquerading as liberal democracy. The fate of the earth hinges upon our heeding voices such as those of Thoreau and Havel and awakening to the meaning of liberal democracy as a comprehensive personal, social, and religious faith.

I would like to pursue this thesis in company of that singular New England democrat, Henry David Thoreau, and our contemporary, Vaclav Havel. The voice of Havel is especially important. Recent events in Eastern Europe have much to teach the world about the liberal democratic faith, and Havel is one of their principal interpreters. His voice is also important to me personally. Havel and I were born the same year, 1936, and in both our cases, the social protest movements of the 1960s formed the crucible of our life commitments.

There are four things we need to consider.

First, we need to understand why the ecological crisis is a crisis in citizenship.

Second, we need to understand why the crisis in citizenship is a crisis in liberal democracy.

Third, we need to understand why the crisis in liberal democracy is a crisis in the revolution of the people.

Fourth, we need to understand why the crisis in the revolution of the people is a crisis of faith.

I

If there is a common cause of global warming, overpopulation, unsustainable economic growth, loss of biodiversity, depletion of natural resources, and the needless suffering of humans and other animals, it is the failure of "we the people," North, South, East, and West, to take moral responsibility for our world. The only possible locus of accountability, and therefore the only proper subject of moral initiative, praise, and blame, is what we do or fail to do as communities of people, our practice of individual and collective self-government. It is easy to lose sight of this obvious and basic fact.

It follows that the ecological crisis is a crisis of citizenship. By definition, a citizen is a member of a community of persons who considers him- or herself morally accountable to the community as a whole. How we think of the meaning of citizenship, its grounds, its bounds, its practices, its purposes, has everything to do with the ecological crisis. "Create citizens, and you will have everything you need," said Rousseau.[4] We also tend to lose sight of this basic fact.

Virtually every reason that we can give for our failure to live well on the earth has its source in some aspect of our consciousness of ourselves as citizens.

If we say a major problem is our failure to face physical and social *limits*, this is rooted, surely, in our failure to conceive of ourselves as a community of self-governing citizens, responsible for the exercise of the moral disciplines necessary to live within our means. Is there any other reason for our failure as Americans to practice what the *New York Times* calls the forbidden "C-word," to consume the major share of the world's oil, and thus to find ourselves at the threshold of war?

If we say a major cause of the ecological crisis is our failure to preserve the *common good*, because the air and water and land, like the welfare of future generations, is a good we all share, this is rooted, surely, in our failure to think and act as a community of citizens, each member of which is responsible for the welfare and future of the community as a whole.

If we say the crisis is due to our failure to share the wealth of the earth, to do *justice*, because without a sense of personal dignity and economic security, people will not care about the environment, and will be driven by necessity to destroy it, this is rooted, surely, in our failure to conceive ourselves a community of equal citizens, each member of which deserves respect and a fair share of the wealth of the society.

If we say the ecological crisis is the result of our failure to *think well* about the facts and imperatives of our situation, this is rooted, surely, in our failure to practice the civic virtues of citizenship, to be informed, to bring the knowledge of experts before the bar of practical reason, to each contribute his or her individual wisdom to the common pool, to deliberate the issues together and decide an effective course of action.

If we say a principal cause of the ecological crisis is our failure to take seriously the particular *places* in which we live, to see how each human society is embedded in a local ecosystem, this is rooted, surely, in our failure to see the actual bounds of the communities to which we belong, to practice what Aldo Leopold calls "the land ethic," which "changes the role of *Homo sapiens* from conqueror of the land-community to plain member and citizen of it."[5]

Finally, if we say the cause of the ecological crisis is our *worldview*, a failure to perceive ourselves as part of the interdependent web of existence, this too is rooted in our idea and practice of citizenship. For how did we come to the idea of the ecological worldview except on the basis of the metaphor of the polis, the "commonwealth," the City of God? As William James asked: "Why may not the world be a sort of republican banquet . . . where all the qualities of being respect one another's personal sacredness, yet sit at the common table of space and time?"[6] If we are not conscious of ourselves as citizens, we will lose the idea of ecology itself.

I conclude that it is in the direction of a profound doctrine of citizenship, one that affirms the free, equal, and mutual responsibility of all persons who sit at the common table of space and time, that our hope lies.

II

Western liberal democracy is often heralded as the most universal, rational, and influential social and political philosophy in the world today. This is the position, for example, of Francis Fukuyama, an employee of the United States State Department, in a hotly debated essay published in *The National Interest* last year. Fukuyama argues that the victory of economic and political liberalism at the end of the twentieth century means the liberal democratic ideal "will govern the material world *in the long run*": "What we may be witnessing is not just the end of the Cold War, or the passing of a particular period of postwar history, but the end of history as such: that is, the end point of Western liberal democracy as the final form of human government."[7]

But in recent years there has also appeared a major post-Marxist critique of liberal democracy that holds it responsible, not for the world's progress, but for most of its problems—including its environmental problems. The environmental movement has been deeply affected by this critique. It divides the movement into factions; and it divides internally as persons many of us who belong to the movement. It leads us sometimes to say and do things in which we do not believe, and to think and feel things we cannot act upon.

There is good reason, therefore, to say that the crisis in citizenship is a crisis in our understanding and practice of liberal democracy. We need to pause for a moment and listen to the two sides of this debate.

On the one hand, it is argued, success in meeting the ecological crisis depends upon the ideas and institutions of liberal democracy.

Where, for example, if it were not for the liberal democratic tradition of *civil and political rights*, would there be an environmental movement in most parts of the world? In all but the most traditional societies, if these rights do not exist, the environment fares poorly.

Nor should we overlook the fact that it is free individuals, often individuals of great personal courage, such as Wangari Maathai, or Chico Mendes, and *free associations* of individuals, such as Greenpeace, or the World Wildlife Fund, that take leadership for the care of the earth.

Moreover, whatever legal guarantees and legislative processes exist in most countries for the protection of the environment originated in the liberal tradition of *limited, constitutional government*—including popular suffrage and an independent judiciary.

I can testify to the ethical power of the liberal democratic tradition through my work on the second edition of the World Conservation Strategy. When those engaged in this project needed to appeal to an international consensus on justice, we turned to the United Nations *Universal Declaration of Human Rights*. When we needed to appeal to an international consensus on environmental ethics, we turned to the United Nations *World Charter for Nature* and its premise that "Every form of life is unique, warranting respect regardless of its worth to man," an obvious extension to the rest of nature of the liberal democratic principle of respect for the intrinsic value of each individual.[8]

Roderick Frazier Nash, in his book *The Rights of Nature*, argues that the history of modern environmental ethics can be best explained by the extension of the liberal concept of "natural rights" to "animals, plants, rocks, and even nature, or the environment, in general."[9] But many other aspects of the tradition are also involved. For example, there is the notion that the individual may experience a radical transformation in self-consciousness through personal encounter with wild nature—the heart of the wilderness movement. And there is the idea that nature may be perfected, and nature and society harmonized, through the practice of the arts and applied sciences—an abiding postulate of humanistic liberalism—and also the democratic ideal of self-reliant and self-sufficient agriculture—so eloquently articulated in recent years by Wendell Berry.

Nor is it possible to understand the contemporary movement toward environmental responsibility apart from two forms of rationality peculiarly associated with liberal democracy—*scientific reason and critical reason*. Tested, empirical knowledge of the natural world sets the base line for all our action. It is in terms of the cosmic evolutionary story told by the modern sciences that we are now rethinking our place and responsibility in the cosmos; it is on the basis of the most recent scientific data that we are rethinking our societal impact upon the biosphere. The history and future of the modern sciences are deeply intertwined with the history and future of the liberal democratic tradition, and the sciences serve a self-correcting function within the tradition.

Equally important is the liberal conviction that all our traditions and personal beliefs must be submitted to searching self-criticism and reconstruction in light of our changing situation and the emergence of new knowledge and insight.

Finally, in spite of the contradictions involved, the economic wealth generated under the aegis of a *free market economy* has liberated countless persons from oppressive toil and scarcity, created leisure to enjoy nature, and in many cases generated the means necessary to protect the environment.

On the other hand, there is widespread disillusionment with liberal democracy as a social philosophy adequate to the challenges of a postindustrial world.

The major substantive criticism is that liberal democracy is an expression of the worldview of modernity, and the modern worldview (Western, Cartesian, Newtonian) is the root of the ecological crisis. Contrary to all that we now know about the interdependence of life, the dominant understanding of liberal democracy remains tied to a kind of "ontological individualism," a belief that the primary reality in the world is autonomous individual subjects existing independently of body, society, and nature and freely pursuing their ends in fierce isolation. The cultural matrix in which this ontology has flourished is the drive of Western technological society toward the domination of nature—the domination not only of nature itself but everything presumed to be associated with it—women, people of color, people who work with their hands, indigenous peoples.

The ideological critique asserts that liberal democracy expresses the interests of the middle classes that came to power in the rise of industrialism and colonialism, that it confirms as "rights" the freedoms this class needs to maintain the separation of the economic sphere from public control. Liberal democracy conveniently overlooks the reality of class conflict, economic inequality, and the regimentation required by industrial modes of production. In effect, it serves as an ideology for the competitive, consumer society that now dominates the world.[10]

For our purposes, what these critiques add up to is an indictment of liberal democracy as a philosophy of citizenship adequate to the ecological age. Liberal democracy does not promote equal, free, and mutual participation of all persons in politics, business, and culture, but instead limits the role of citizens to choosing which elite will govern them. Everything else is left to what is called "society": that freewheeling, amorphous region of mass public opinion, voluntary association, and individual pursuit of economic gain. As Vermont social ecologist Murray Bookchin describes it, "A 'good citizen' is one who obeys the laws, pays taxes, votes ritualistically for preselected candidates, and 'minds his or her own business.' "[11] Liberal democracy is not, in other words, democratic, in the essential and enduring sense of democracy as "the power of the people to govern themselves."

Furthermore, liberal democracy has no idea of the common good. Nor does it recognize obligations that arise from the fact of our membership in the human and natural community. It works on behalf of our most immediate personal interests—not on behalf of the welfare of future generations or those who live beyond our shores. It promotes neither a deep commitment to social justice nor a profound sense of our dependence upon, and gratitude for, the rest of nature. It has, in other words, no "transcendent reference." It has broken the bonds of shared ritual and belief that once united human communities with their environments and shattered the great unifying "wholes" of our existence—nature and God.

In sum, liberal democracy is a philosophy for the pursuit of private happiness, not for public citizenship.

If we accept the burden of this critique, as I do, we are left with a dilemma. To continue to support liberal democratic values and institutions is to undercut the positive understanding of citizenship necessary to address the global ecological crisis. But to reject liberal democracy is to cut ourselves off from the most successful democratic tradition in the world today, and its precious inheritance of human rights, reason, and government under law—an inheritance more precious, we must add, with every passing day.

III

There is, however, another view of liberal democracy, and another experience of it—one based on a strong, communal understanding of citizenship and our interdependence with the rest of life, which at the same time affirms the crucial liberal principles of individual freedom and reason. We may call this radical, revolutionary, or prophetic liberal democracy, to distinguish it from the dominant understanding, so heavily influenced by seventeenth-century Enlightenment philosophy, Newtonian science, and economic liberalism. I believe that it is on behalf of this alternative democratic ideal and experience that Thoreau and Havel speak.

Vaclav Havel is careful in his use of words. As a writer, he is terribly conscious of their "mysteriously ambiguous power." "The selfsame word," he writes, "can at one time be the cornerstone of peace, while at another, machine-gun fire resounds in its every syllable."[12] "Liberty" and "democracy" are such words. Precisely because they are such powerful and important words, they are capable of the most extreme and demonic perversions.

Havel's fear is that democracy will suffer the fate of socialism and be degraded to an "uninhabitable fiction." He is therefore cautious about identifying what he means by "democracy" with the word as customarily used and has coined such terms as "anti-political politics" and "post-democracy" to describe what actually happened in Czechoslovakia.

Havel accepts most of the critique of the prevailing understanding of liberal democracy outlined earlier. In his view, for example, what Western "parliamentary democracy" calls "freedom" is an "escape from the sphere of public activity" into the sphere of merely private life. The "post-totalitarian" societies of Eastern Europe are an exaggerated form of the Western consumer and industrial way of life, and therefore a warning to the West of its own latent tendencies. "Instead of free economic decision-sharing, free participation in political life, and free intellectual advancement," he writes, all that the citizen of either system is "actually offered is a chance freely to choose which washing machine or refrigerator he wants to buy."[13]

At the same time, Havel strongly affirms fundamental liberal principles, such as the separation between the center of political power and the center of truth. Havel speaks as a true liberal when he writes that one cannot abdicate "one's own reason, conscience, and responsibility" to any "higher authority,"[14] and "the only possible place to begin is with myself. . . ."[15]

He is able to reconcile these two positions because he holds that individual freedom and interdependent community require one another. True personal freedom is based on solidarity and is fulfilled in public duty.

This is clear in his discussion of what it meant to be a "dissident." Western journalists by and large misunderstand this term—which they invented. In Western eyes, dissident means a special category of individuals who are brave enough to stand alone *against* the system. But this contradicts the real nature of the dissident attitude which "stands or falls on its interests in others," "on what ails society as a whole," and on the principle that the rights and freedoms of citizens are indivisible. "It is truly a cruel paradox that the more some citizens stand up in defense of other citizens," he writes, "the more they are labelled with a word that in effect separates them from those 'other citizens.'"[16]

Havel was equally upset when Westerners asked dissidents, "What can we do for you?" The implication of the question is that the issue at stake is the welfare of the dissidents, not "the salvation of us all, of myself and my interlocutor equally. . . . Are not my dim prospects or, conversely, my hopes his dim prospects and hopes as well? Was not my arrest an attack on him and the deceptions to which he is subjected an attack on me as well? Is not the destruction of humans in Prague a destruction of all humans?"[17]

He could have quoted Thoreau who in "Civil Disobedience" declared: "Under a government which imprisons any unjustly, the true place for a just man is also a prison."[18]

In Havel's view, there are never "mere" individuals or "mere" communities but always individuals-*in*-community, each individual having by nature both the freedom and responsibility to make a unique contribution to the good of the whole. Thoreau gave voice to this same insight when he wrote in his *Journal*, on 26 March 1842: "I will sift the sunbeams for the public good. I know no riches I would keep back. I have no private good, unless it be my peculiar ability to serve the public."[19]

It is therefore vital to the welfare of the community that it recognize as "rights" each person's capacity for freedom, reason, and responsibility, and that it provide the material and cultural conditions to make these rights meaningful. But it is the community of individuals that holds these rights and obligations, not single individuals only, and therefore the community must also be enfranchised as a self-governing community. We must always speak, therefore, of individual *and* community, rights *and* duties, freedom *and* equality, plurality *and* solidarity, *liberalism* and *democracy*.

In summary, the radical principle of liberal democracy, as Havel and Thoreau affirm it, is rooted in the principle of individuals in community, which is the principle of citizenship, which is the principle of ecology. This is not a Western, or modern, but a universal ideal.[20]

But this means that beneath the crisis in the prevailing understanding of liberal democracy as a philosophy of citizenship adequate to the ecological crisis, there is a deeper crisis, the crisis among us, the people. It is our failure to fulfill the revolutionary promise of liberal democracy in the modern world.

For when we trace the historical origins of prophetic liberal democracy to their most proximate historical sources, they lie in those periods and movements when "we the people" asserted our sovereignty and interpreted "government by the consent of the governed" to mean actual self-government. "The government is only the mode which the people have chosen to execute their will," Thoreau wrote in "Civil Disobedience." "Even voting *for the right*," he went on, "is *doing* nothing for it. It is only expressing to men feebly your desire that it should prevail."[21] In principle, at least, "we the people" wrote and adopted the constitutions of the free nations of the world; and "we the peoples of the united nations" wrote the *Charter of the United Nations*.[22]

But who now governs the world? Or any major nation within it?

The events in Eastern Europe were "liberal democratic revolutions" because the peoples of the region asserted their fundamental right to rule the commons, to be, in the full sense of the word, "citizens." It was no accident that the Czech dissident movement called itself the Civil Forum, and that opposition groups in East Germany called themselves "citizens' initiatives."

Havel is emphatic about this point. The failure of each person to take personal responsibility for what he calls "those matters that concern us all" was the principal cause of the slide of Eastern Europe into "post-totalitarianism."[23] "I feel," he writes, "along with my fellow-citizens—a sense of culpability for our former reprehensible passivity. . . ."[24] Only when the people began to act as citizens "from below," to participate in the creation of a parallel polis, "the *locus* of a renewed responsibility for the whole and to the whole," did matters begin to change.[25] The "parallel polis" was the foundation on which the "main polis," the society itself, was ultimately transformed. This was not only a political revolution, in the narrow sense of the term, but an ecological revolution, for as Havel points out, the ecological movement is citizenship, or "responsibility to and for the whole" in its most universal form.

The crisis in liberal democracy is therefore the crisis of *our* revolution, the moral and social revolutions that we, the people, have repeatedly made in the course of human history, from the time of the founding of the great civic and religious traditions of the ancient world to the founding of modern nations, but whose universal purpose to liberate life from oppression through responsible self-government has failed. The question this raises is why we failed and how we may ultimately prevail.

IV

Of all the reasons that might be given for our failure, I believe there is only one, finally, that goes to the root of the matter: it is a failure of faith. This was Vaclav Havel's message to the United States Congress in February of 1990.

While acknowledging that Eastern Europe has much to learn from the West about free elections and free markets, he insisted that the West also has much to learn from the East: "a special capacity to look, from time to time, somewhat further than someone who has not undergone [our] bitter experience."[26]

Havel then spelled out what the West had to learn: that it is a long way from realizing the ideal of democracy, that it is "still under the sway of the destructive and vain belief that man is the pinnacle of creation, and not just a part of it, and that therefore everything is permitted," that it has yet to understand that the salvation of the world lies in a "global revolution in human consciousness," not in economic growth, that the locus of this transformation is the conscience of each and every citizen, worker as well as intellectual, and that this is so because conscience is the mediator of our responsibility to "the order of Being."[27]

What Havel in effect told the American people was that they have lost their faith; and he concluded by noting that what they have lost is of no small moment in world history. For their faith wrote the Declaration of Independence, the Bill of Rights, and the Constitution, and it still has power to "inspire us to be citizens." Havel was retrieving the faith of the eighteenth-century democratic revolutions when such phrases as "we the people," "consent of the governed," "freedom of conscience," "inalienable rights," and "the laws of nature and Nature's God" were living and self-evident truths.

This was the faith that many in Eastern Europe rediscovered in their "bitter experience" since the Second Great War, and it is the truth they now have to share with the rest of the world.

It is striking that the slogan of the Czechoslovakian revolution was the Hussite conviction, "the truth shall prevail"; that these are the words Thomas Jefferson echoed when he wrote in the "Act for Establishing Religious Freedom in Virginia": "that truth [which is universal] is great," and if left to herself, "will prevail"; and that when Havel tried to put his finger on the spiritual center of the Czechoslovakian revolution he used the phrase, "living in the truth."

Havel explains the meaning of "living in the truth" in his essay, "Politics and Conscience." He opens with a personal experience:

> As a boy, I lived for some time in the country and I clearly remember an experience from those days: I used to walk to school in a nearby village along a cart track through the fields and, on the way, see on the horizon a huge smokestack of some hurriedly built factory, in all likelihood in the service of war. It spewed dense brown smoke and scattered it across the sky. Each time I saw it, I had an intense sense of something profoundly wrong, of humans soiling the heavens. I have no idea whether there was something like a science of ecology in those days; if there was, I certainly knew nothing of it. Still that "soiling the heavens" offended me spontaneously. It seemed to me that, in it, humans are guilty of something, that they destroy something important, arbitrarily disrupting the natural order of things, and that such things cannot go unpunished.[28]

Interpreting this event, Havel says that the sense of good and evil he so strongly felt that day was his personal experience of the Absolute. It was an experience of a reality that youth takes for granted but from which most modern adults are alienated. Yet there is nothing more directly accessible to us than this world of our "lived experience"—this world prior to the split of subject and object, fact and value, spirit and nature, this world with its morning and its evening, its up and its down, where, he writes, "good and evil, beauty and ugliness, near and far, duty and work, still mean something living and definite." This is the realm of ordinary human experience and of the human conscience. It is "the realm of our induplicable, inalienable and non-transferable joy and pain, a world in which, through which, and for which we are somehow answerable, a world of personal responsibility."[29]

The faith of prophetic liberal democracy is that this immediate and absolute world of ordinary human experience, this world that gathers us up while maintaining our plurality, this world we share with each other and all other living things on this earth, is the sphere of ultimacy. There is no greater truth and no greater accountability. The only proper response before it is profound humility and respect.

This is the world that the Eastern European "dissidents" rediscovered under the bitter conditions of post-totalitarianism. With every other prop to individual dignity removed, they were forced to find their salvation in the only place it can be found: in their personal experience of the irreducible unity and goodness of life and in their individual conscience.

What Havel is articulating in a contemporary and powerful form is the eighteenth-century revolutionary belief in "natural religion." This faith held that whatever we need to know of the Creator is manifest in the Creation. The moral, spiritual, and scientific truths necessary for human beings to live well together and in harmony with the rest of nature are accessible to all—"the truth shall prevail." They are available in the natural law present to our ordinary human senses and reason, and in the "higher" law accessible to our God-given capacity for moral judgment. In this way, experience and conscience become the means to truth "of, for, and by the people." Truth is personally and uniquely apprehended but publicly shared and tested.

Those of us who are privileged to live in liberal democratic societies take for granted the radical change in human history that this faith entailed. It was a reversal in the historic understanding of the relation between natural revelation and special revelation, a reversal with profound consequences for the relation of religion and politics. Prior to the seventeenth and eighteenth centuries, it was generally presumed that while natural revelation may provide a common ground for rational intimations of a divine being, it was only in the special revelation of a particular revealed religion that these intimations could lead to a real faith in the true God. This meant an established religion or theocracy, since supremacy was assumed to be the proper relation of special revelation to the rest of life.

Prophetic liberal democracy reversed this: what is morally, spiritually, and politically prior is natural revelation, for it is inclusive of all special revelations. Moreover, natural revelation is not only a rational intimation of our common Creator and our common creature-hood here below, it is a deeply felt intimation of our common redemption as well. We will be saved, as we are born, and as we must live, together, on this earth. The fate of our common creation encompasses and transcends the fate of our common humanity.

Thus it is the liberal democratic faith that ultimately grounds our rights of individual conscience, freedom of thought, speech, and worship. These rights are instituted to protect each person's unique exercise of moral self-government founded, in turn, on each person's unique capacity for natural revelation—for the private discernment of public truth; because it is precisely out of our plurality, not our uniformity, that "the truth shall prevail."

This holds true for traditions as well as individuals. Any particular religious tradition rests upon a particular historical revelation which is its unique grasp of universal truth. We can never reach the universal except through the particular. But only as we plumb the depths of the particular, and share the particular truths we have found, will we understand the Whole. Again, the only Absolute presupposed is that all the peoples and religions of the world share with one another and the earth a common origin, life, and destiny.

It follows from this structure of belief that "man is not God," "woman is not God," no one of us, no group of us, not all of us. The truth, and *only* the truth, shall prevail. Here is the basis for the separation, in principle, of all political and worldly power from the truth. Public life shall not be controlled by any special, or privileged, revelation by any race, class, religion, or gender. As Havel writes, the "principle of exclusion" is the basis of all evil.[30]

Sometimes it is said that the liberal democratic faith does not require belief in God; one may be "anything one chooses," an atheist, a humanist, a theist. This is true if by it we mean that no *particular* idea of God is required, but far from true if we mean that no higher order or authority, no Absolute is presupposed in human experience. As Thoreau wrote, "This is the only way, we say; but there are as many ways as there can be drawn radii from one centre."[31]

This, however, is a minimal understanding of the liberal democratic faith. The real God of liberal democracy is a God of overflowing fertility, growth, plenitude, and abundance. It is what Havel tries to describe when he writes: "Life rebels against all uniformity and leveling; its aim is not sameness, but variety, the restlessness of transcendence, the adventure of novelty and rebellion against the status quo. An essential condition of its enhancement is the secret constantly made manifest."[32] And it is what Thoreau also tried to describe when he wrote: "I see, smell, taste, hear, feel, that everlasting Something to which we are allied, at once our maker, our abode, our destiny, our very Selves . . . the actual glory of the universe."[33]

Since no individual, indeed, no people, will ever experience more than an infinitesimal part of this glory, each person and each religion should be eager to learn from every other, and to engage in the most precious ritual in the religion of democracy, public dialogue and mutual persuasion.

The truth *shall* prevail!

To awaken to the liberal democratic faith is to awaken to Spirit in Nature. It is to become a political and ecological revolutionary. For if we hold this faith, "we the people" have grounds to trust the ultimate identity between our spirits and the Absolute Spirit, and this means we have grounds to think for ourselves, to trust our own experience, to trust one another, to trust the world, to believe that what we find good will be preserved in the order of things. And this means *all* of us, including the *least* of us, and it means *all* of nature, including the *least thing* in nature.

To lose this faith is to cut the nerve of liberal democracy, to remove the motivation to live responsibly under the difficult, often tragic, conditions of life as it is given to us. In Thoreau's judgment, "It is hard to be a good citizen of the world in any *great* sense. . . ."[34] To doubt that Spirit infuses

Nature is to doubt that citizenship is possible. It is to reduce personal experience and responsibility to an absurdity. It is to lapse into a world of isolated selves and to sever human experience from nature.

Havel believes that the reason we have failed to carry forward the revolutionary promise of liberal democracy is that we have betrayed this faith. In effect, we have violated the common ground of our being.

We must then forgive one another and renew the covenant. For the fate of the earth depends on our reconstituting the natural world as the true terrain of citizenship. This means pledging ourselves, in Havel's words, to

> . . . draw our standards from our natural world, heedless of ridicule, and reaffirm its denied validity. We must honour with the humility of the wise the bounds of that natural world and the mystery which lies beyond them, admitting that there is something in the order of being which evidently exceeds all our competence; relating ever again to the absolute horizon of our existence which, if we but will, we shall constantly discover and experience.[35]

This is why, he concludes, our hope lies with those "single, seemingly powerless" persons who dare "to cry out the word of truth and to stand behind it."[36]

NOTES

1. Henry David Thoreau, *Natural History Essays* (Salt Lake City: Gibb-Smith, 1980), 134.
2. Vaclav Havel, "The Future of Central Europe," Speech to the Polish Sejm and Senate, 21 January 1990, *New York Review of Books*, 29 March 1990, 19.
3. Henry David Thoreau, "Life Without Principle," in *Anti-Slavery and Reform Papers* (Montreal: Harvest House, 1963), 146.
4. Jean-Jacques Rousseau, "A Discourse on Political Economy," in *Social Contract and Discourses* (London: Dent, 1913), 251.
5. Aldo Leopold, *A Sand County Almanac and Sketches Here and There* (London: Oxford University Press, 1949), 204.
6. William James, *The Will to Believe* (New York: Dover Publications, 1956), 270.
7. Francis Fukuyama, "The End of History?" *The National Interest* 16 (Summer 1989): 3.

8. Wolfgang E. Burhenne and Will A. Irwin, *The World Charter for Nature: A Background Paper* (Berlin: Erich Schmidt Verlag GmbH, 1983), 9.
9. Roderick Frazier Nash, *The Rights of Nature: A History of Environmental Ethics* (Madison: The University of Wisconsin Press, 1989), 4.
10. For these critiques, see Anthony Arblaster, *The Rise and Decline of Western Liberalism* (New York: Basil Blackwell, 1984); Alfred W. Crosby, *Ecological Imperialism: The Biological Expansion of Europe, 900–1900* (New York: Cambridge University Press, 1986); Carolyn Merchant, *The Death of Nature: Women, Ecology, and the Scientific Revolution* (New York: Harper and Row, 1980); Michael Sandel, ed., *Liberalism and Its Critics* (New York: New York University Press, 1974).
11. Murray Bookchin, *The Rise of Urbanization and the Decline of Citizenship* (San Francisco: Sierra Club Books, 1987), 9.
12. Vaclav Havel, "Words on Words," Acceptance Speech for the Peace Prize of the German Booksellers Association, 15 October 1989, *New York Review of Books* 36 (18 January 1990): 5.
13. Vaclav Havel, *Living in the Truth*, ed. Jan Vladislav (Boston: Faber and Faber, 1986), 13.
14. Ibid., 39.
15. Quoted from Vaclav Havel, *Letters to Olga,* in Richard L. Stanger, "Vaclav Havel: Heir to a Spiritual Legacy," *The Christian Century* (11 April 1990), 369. The passage runs: "But who should begin? Who should break this vicious circle? The only possible place to begin is with myself . . . it is I who must begin. . . . For the hope opened up in my heart by this turning toward Being has opened my eyes as well. . . . Whether all is really lost or not depends entirely on whether I am lost."
16. Havel, *Living in the Truth*, 79–80.
17. Ibid., 149.
18. Henry David Thoreau, *Journal of Henry D. Thoreau*, ed. Sherman Paul (Boston: Houghton Mifflin, 1957), 245.
19. Henry David Thoreau, *Journal of Henry D. Thoreau*, ed. Bradford Torrey and Francis H. Allen (New York: Dover Publications, 1962), vol. 1, 106.
20. John Dewey, who in many respects anticipated the contemporary critique of liberal democracy, wrote in *The Public and Its Problems* ([New York: Holt, 1927], 148): "Regarded as an idea, democracy is not an alternative to other principles of associated life. It is the idea of community life itself. It is an ideal in the only intelligible sense of an ideal: namely, the tendency and movement of some thing which exists carried to its final limit, viewed as completed, perfected. Wherever there is conjoint activity whose consequences are appreciated as good by all singular persons who take part in it, and where the realization of the good is such as to effect an energetic desire and effort to sustain it in being just because it is a good shared by all, there is in so far a community. The

clear consciousness of a communal life, in all its implications, constitutes the idea of democracy."

21. Henry David Thoreau, *Walden and Civil Disobedience*, ed. Sherman Paul (Boston: Houghton Mifflin, 1957), 235, 240.
22. Burns H. Weston, Richard A. Falk, and Anthony A. D'Amato, eds., *Basic Documents in International Law and World Order* (St. Paul, Minn.: West Publishing, 1980), 7.
23. Havel, *Living in the Truth*, xiii.
24. Vaclav Havel, "Help the Soviet Union on its Road to Democracy: Consciousness Precedes Being," Speech delivered to the Joint Session of Congress, Washington, D.C., 21 February 1990, *Vital Speeches of the Day* (56), 329.
25. Havel, *Living in the Truth*, 112.
26. Havel, "Help the Soviet Union on its Road to Democracy," 330.
27. Ibid.
28. Havel, *Living in the Truth*, 136.
29. Ibid., 137.
30. Ibid., 6.
31. Henry David Thoreau, *Walden*, ed: J. Lyndon Shanley (Princeton: Princeton University Press, 1971), 11.
32. Havel, *Living in the Truth*, 23.
33. Henry David Thoreau, *The Illustrated A Week on the Concord and Merrimack Rivers*, ed. Carl Hovde, William Howarth, and Elizabeth Witherell (Princeton: Princeton University Press, 1983), 173–74.
34. Henry David Thoreau, *Journal of Henry D. Thoreau*, ed. Bradford Torrey and Francis H. Allen (New York: Dover Publications, 1962), vol. 1, 106. Emphasis added.
35. Havel, *Living in the Truth*, 153.
36. Ibid., 156.

CHAPTER 5

Islam and the Environmental Crisis

Seyyed Hossein Nasr

Fig. 5. As Louis Agassiz Fuertes's (1874–1927) falcon hunches over her prey, we experience the vital integrity of the biosphere: the teal is about to become a peregrine, the peregrine, a teal. Wings interfolded, the birds are one. Peregrine falcon, Falco peregrinus with Green-winged Teal, Anas carolinensis *(no date) also offers hope in an era when there is so much environmental loss. By about 1960, peregrines were extinct east of the Mississippi. Now, with human assistance, populations of these spectacular aerialists have been reestablished in many parts of the East. (Library, The Academy of Natural Sciences of Philadelphia.)*

Islam is a monotheistic religion with roots in ancient Judaism and early Christianity as well as in Arab culture. Despite these connections, however, the understanding and appreciation of the Islamic tradition in Western societies has been relatively limited. Since the 1960s there has been a dramatic growth of interest among Westerners in non-Western religions, but Asian traditions like Buddhism have received much more attention than Islam. In part due to international economic and political developments as well as a deepening awareness of the treasures to be found in Islamic spiritual teachings, this is changing. Programs in Islamic studies are sprouting in American colleges and universities, and the 1990s may well be a decade when the dialogue between Western religions and Islam is fully engaged. One force bringing these traditions into closer relations is the emergence of a shared ethical concern about the deterioration of the environment and its cultural causes.

The work of Professor Seyyed Hossein Nasr has helped to lay the foundations for a productive dialogue between Islam and Western culture. He has devoted his life to the study of Islamic philosophy and spirituality and to exploring the relation of Islam to modern culture. In books like Islam and the Plight of Modern Man *(1975), he has endeavored to present the significance of Islam to Western men and women searching for wholeness and a spiritual center amidst the moral and religious confusion of secular urban/industrial society. He also seeks to address "the modernized Muslim," who has become concerned to preserve Islamic civilization in the face of the "corrosive" influence of Western secularism and materialism. Professor Nasr could clearly see the environmental crisis unfolding in the 1960s, and he was among the first to recognize it as a spiritual crisis and to call upon the religions to become involved.*

In his essay Hossein Nasr offers a critique of Western humanism and secular science from the perspective of the Islamic tradition. He argues that the environmental crisis arose when society forgot God and lost an appreciation of, and a language for, the sacred quality of nature. Profoundly influenced by Sufism, which forms the inner and mystical dimension of Islam, Nasr believes that only a spiritual rebirth of the individual, involving a reawakening to the Divine Center, will bring about an enduring solution to our current problems. Reflecting on the situation in the Islamic world, he asserts that Islamic peoples have often been misguided by Western industrial civilization, leading to the degradation of the environment. However, the destructive trends can be reversed by a recovery of the traditional Islamic understanding of the interrelation of God, humanity, and nature, which involves a sacramental view of the physical universe. God encompasses and permeates nature, and nature is a great book, "the cosmic Quran," in which the divine beauty and truth are symbolized. The Islamic world, he contends, can find the foundations for an ethic adequate to an age of environmental crisis in this theological vision of the sacredness of nature and in the revealed truth of the Sharīʿah, *the Divine Law.*

O LORD, SHOW US THINGS AS THEY REALLY ARE.
—*Saying (*ḥadīth*) of the Prophet*

WHEN one looks at the Islamic world today, one sees blatant signs of the environmental crisis in nearly every country, from the air pollution of Cairo and Tehran to the erosion of the hills of Yemen to the deforestation of many areas of Malaysia and Bangladesh. Environmental problems seem to be present everywhere especially in urban centers and also in many parts of the countryside to a degree that one cannot distinguish the Islamic world from most other areas of the globe as far as acute environmental problems are concerned. If one were to study the situation only superficially, one could in fact claim, in the light of present-day conditions, that the Islamic view of nature could not have been different from that of the modern West which first thrust the environmental crisis upon the whole of mankind. But a deeper look will reveal an Islamic view of the environment very different from what has been prevalent in the West during the past few centuries. If that view has now become partly hidden, it is because of the onslaught of Western civilization since the eighteenth century and the destruction of much of Islamic civilization, due to both external and internal factors, although the Islamic religion itself has continued to flourish and remains strong.

In fact, the Islamic world is not totally Islamic today, and much that is Islamic lies hidden behind the cover of Western cultural, scientific, and technological ideas and practices emulated and aped to various degrees of perfection, or rather one should say of imperfection, by Muslims during the past century and a half. The Islamic attitude toward the natural environment is no more manifest than the Buddhist one in Japan or Taoist one in China, all as a result of the onslaught upon these lands of a secular science based upon power and domination over nature and a technology which devours the natural world with no respect for the equilibrium of nature, a science and technology of Western origin which have now become nearly global.

Despite this situation, however, Islam continues to live as a powerful religious and spiritual force, and its view of nature and the natural environment still has a hold upon the mind and soul of its adherents especially in less modernized areas and also in some of the deeper attitudes toward nature. The role of this survival of the traditional view of nature can be seen in the refusal of Islamic society to surrender completely to the dicta of the machine despite the attempt of leaders of that world to introduce Western technology as much as and as soon as possible. This view is, therefore, significant for a global consideration of the environmental problem not only because of its innate value but also because of its continuous influence upon Muslims who comprise a fifth of the world's population.

The Islamic view of the natural environment is furthermore of significance for the West itself, since Islam shares with the West a religion of the Abrahamic family and the Greek heritage, which played a major role in the history of both Western and Islamic science—in Western science mostly through the agency of Islamic science. The Islamic view of nature presents a precious reminder of a perspective mostly lost in the West today. It is based upon the sacred quality of nature in a universe created and sustained by the One God of Abraham to whom Jews and Christians also bow in prayer.

The Islamic view of the natural order and the environment, as everything else that is Islamic, has its roots in the Quran, the very Word of God, which is the central theophany of Islam.[1] The message of the Quran is in a sense a return to the primordial message of God to man. It addresses what is primordial in the inner nature of men and women; hence Islam is called the primordial religion (*al-dīn al-ḥanīf*).[2] As the "Primordial Scripture," the Quran addresses not only men and women but the whole of the cosmos. In a sense, nature participates in the Quranic revelation. Certain verses of the Quran address natural forms as well as human beings, while God takes nonhuman members of His creation, such as plants and animals, the sun and the stars to witness in certain other verses. The Quran does not draw a clear line of demarcation between the natural and the supernatural nor between the world of man and that of nature. The soul which is nourished and sustained by the Quran does not regard the world of nature as its natural enemy to be conquered and subdued but as an integral part of its religious universe sharing in its earthly life and in a sense even its ultimate destiny.

The cosmic dimension of the Quran became elaborated over the centuries by many Muslim sages who referred to the cosmic or ontological Quran (*al-Qurʾān al-takwīnī*) as distinct from and complementing the composed or "written" Quran (*al-Qurʾān al-tadwīnī*).[3] They saw upon the face of every creature letters and words from the pages of the cosmic Quran which only the sage can read. They remained fully aware of the fact that the Quran refers to the phenomena of nature and events within the soul of man as *āyāt* (literally signs or symbols), a term that is also used for the verses of the Quran.[4] They read the cosmic book, its chapters and verses, and saw the phenomena of nature as "signs" of the Author of the book of nature. For them the forms of nature were literally *āyāt Allāh*, *vestigia Dei*, a concept that was certainly known to the traditional West before symbols were turned into brute facts with the advent of rationalism and before the modern West set out to create a science to dominate over nature rather than to gain wisdom and joy from the contemplation of its forms.

The Quran depicts nature as being ultimately a theophany which both veils and reveals God. The forms of nature are so many "masques" which hide various Divine Qualities while also revealing these same Qualities for those whose inner eye has not become blinded by the concupiscent ego and the centripetal tendencies of the passionate soul.

In an even deeper sense, it can be claimed that according to the Islamic perspective God Himself *is* the ultimate environment which surrounds and encompasses humanity. It is of the utmost significance that in the Quran God is said to be the All-Encompassing (*Muḥīṭ*), as in the verse, "But to God belong all things in the heavens and on the earth: And He it is who encompasseth (*muḥīṭ*) all things" (IV:126); and that the term *muḥīṭ* also means environment.[5] In reality, humans are immersed in the Divine *Muḥīṭ* and are only unaware of it because of their own forgetfulness and negligence (*ghaflah*), which is the underlying sin of the soul to be overcome by remembrance (*dhikr*). To remember God is to see Him everywhere and to experience His reality as *al-Muḥīṭ*. The environmental crisis may in fact be said to have been caused by the human refusal to see God as the real "environment" which surrounds us and nourishes our life. The destruction of the environment is the result of the modern attempt to view the natural environment as an ontologically independent order of reality, divorced from the Divine Environment without whose liberating grace it becomes stifled and dies. To remember God as *al-Muḥīṭ* is to remain aware of the sacred quality of nature, the reality of natural phenomena as signs (*āyāt*) of God and the presence of the natural environment as an ambience permeated by the Divine Presence of that Reality which alone is the ultimate "environment" from which we issue and to which we return.

The traditional Islamic view of the natural environment is based on this inextricable and permanent relation between what are today called the human and natural environments and the Divine Environment which sustains and permeates them. The Quran alludes in many verses to the unmanifested and the manifested worlds (*ʿālam al-ghayb waʾl-shahādah*). The visible or manifested world is not an independent order of reality but a manifestation of a vastly greater world which transcends it and from which it issues. The visible world is like what one can observe around a campfire during a dark desert night. The visible gradually recedes into the vast invisible which surrounds it and for which the invisible is the veritable environment. Not only is the invisible an infinite ocean compared to which the visible is like a speck of dust, but the invisible permeates the visible itself. It is in this way that the Divine Environment, the Spirit, permeates the world of nature and of normal humanity, nourishing and sustaining them, being at once the origin (*al-mabda*ʾ) and entelechy or end (*al-maʿād*) of the manifested order.[6]

As a result of this view of nature as delineated in the Quran and accentuated by the sayings (*Ḥadīth*) and wonts (*Sunnah*) of the Prophet, the traditional Muslim always harbored a great love for nature which is a reflection of paradisal realities here below. This love is reflected not only in Arabic, Persian, and Turkish literature—not to speak of the literatures of other Islamic peoples from Swahili to Malay—but also in Islamic religious thought where no clear distinction is made between what Western theology has come to call the natural and the supernatural. This love is also reflected in many pages of the works of Islamic philosophers but finds its most profound and also universal expression in Sufi poetry. It was the Persian poet Saʿdī who composed the famous verse: "I am joyous with the cosmos for the cosmos receives its joy from Him. I love the whole world, for the world belongs to Him." It was another Sufi, this time the great folk poet of the Turkish language, Yunus Emre, who heard the invocation of God's Blessed Name in the sound of flowing streams which brought a recollection of paradisal realities, and so he sang,

> The rivers all in Paradise
> Flow with the word Allah, Allah,
> And ev'ry loving nightingale
> He sings and sings Allah, Allah.[7]

The Muslim contemplatives and mystics have loved nature with such intensity because they have been able to hear the prayer of all creatures of the natural world to God.[8] According to the Quran, "Nothing is, that does not proclaim His praise" (XVII: 44). This praise, which is also the prayer of all things, is the root of their very existence. Fallen man, who has forgotten God, has become deaf to this ubiquitous prayer as a result of this very act of forgetfulness. The sage on the contrary, lives in remembrance of God (*dhikr Allāh*) and as a result hears the prayers of flowers as they turn toward the sun and streams as they descend from hills toward the sea. The sage's prayer has in fact become one with the prayer of the birds and the trees, of the mountains and the stars. He prays with them and they with him, and in contemplating their forms not only as outwardness but as theophanies or as "signs of God," the sage is strengthened in recollection and remembrance of the One. They in turn draw an invisible sustenance from the human being who is open to the grace emanating from the realm of the Spirit and who fulfills his or her role as the *pontifex*, the bridge between heaven and earth.

This contemplative attitude toward nature and love for it is, of course, reserved at its highest level for the few who have realized the full possibilities of being human, but throughout the centuries it has percolated into the Islamic community as a whole. Traditional Islamic society has always been noted for its harmonious relation with the natural environment and love for nature to the extent that many a Christian critic of Islam has accused Muslims of being naturalistic and Islam of being devoid of the grace which is usually so trenchantly separated and distinguished from nature in the mainstream of Christian theology.

This Islamic love of nature as manifesting the "signs of God" and being impregnated by Divine Presence must not be confused with naturalism as understood in Western philosophy and theology. Christianity, having been forced to combat the cosmolatry and rationalism of the ancient Mediterranean world, branded as naturalism both the illegitimate nature worship of the decadent forms of Greek and Roman religion and the very different concern and love for nature of the northern Europeans such as the Celts. This love nevertheless survived in a marginal manner after the Christianization of Europe, as one can see in the works of Hildegard of Bingen or in early medieval Irish poems pertaining to nature. Now, one must never forget that the Islamic love for nature has nothing to do with the naturalism anathematized by the Christian church fathers. Rather, it is much closer to the nature poetry of the Irish monks and the addresses to the sun and the moon by the patron saint of ecology, St. Francis of Assisi. Or perhaps it should be said that it is he who among all the great medieval saints is the closest to the Islamic perspective as far as the love of nature is concerned. In any case, the Islamic love of nature and the natural environment and the Islamic emphasis upon the role of nature as a means to gain access to God's wisdom as manifested in His creation do not in any sense imply the negation of transcendence or the neglect of the archetypal realities. On the contrary, on the highest level, it means to understand fully the Quranic verse, "Whithersoever ye turn, there is the Face of God" (II:115). It means, therefore, to see God everywhere and to be fully aware of the Divine Environment which surrounds and permeates both the world of nature and the ambience of humanity.

The Islamic teachings concerning nature and the environment cannot be fully appreciated without dealing with the Islamic understanding of humanity. Man, who has always been viewed in various traditional religions as the custodian of nature, has now become its destroyer. Man has changed roles, thanks to modern civilization, from the being who had descended from Heaven and lived in harmony with the earth to a creature who is thought to have ascended from below and who has now become the earth's most deadly predator and exterminator. Islam considers man as God's vice-gerent (*al-khalīfah*) on earth, and the Quran asserts explicitly, "I am setting on the earth a vice-gerent (*khalīfah*)" (II:30). This quality of vice-gerency is, moreover, complemented by that of servantship (*al-ʿubūdiyyah*) toward God. Man is God's servant (*ʿabd Allāh*) and must obey Him accordingly. As *ʿabd Allāh*, man must be passive toward God and a recipient of the grace that flows from the world above. As *khalīfat Allāh*, man must be active in the world, sustaining cosmic harmony and disseminating the grace for which humans are the channel as a result of their being the central creatures in the terrestrial order.[9]

In the same way that God sustains and cares for the world, humankind as His vice-gerent must nurture and care for the ambience in which they play the central role. They cannot neglect the care of the natural world without betraying that "trust" (*al-amānah*) which they accepted when they bore witness to God's lordship in the pre-eternal covenant (*al-*

mithāg), to which the Quran refers in the famous verse, "Am I not your Lord? They [that is, members of humanity] uttered, yea we bear witness" (VII:172).

To be human is to be aware of the responsibility which the state of vice-gerency entails. Even when in the Quran it is stated that God has "subjected" (*sakhkhara*) nature to man, as in the verse "hast thou not seen how God has subjected to you all that is in the earth" (XXII:65), this does not mean the ordinary conquest of nature as claimed by so many modern Muslims thirsty for the power which modern science bestows upon man. Rather, by it is meant the dominion over things which humans are allowed to exercise only on the condition that it be according to God's laws and precisely because they are God's vice-gerent on earth, being therefore given a power which ultimately belongs to God alone and not to them as merely creatures born to journey through this earthly life and to return to God at the moment of death.

That is also why nothing is more dangerous for the natural environment than the practice of the power of vice-gerency by a humanity which no longer accepts its place as God's servant, obedient to His commands and laws. There is no more dangerous creature on earth than a *khalīfat Allāh* who no longer accepts the role of *ʿabd Allāh* and who therefore does not give allegiance to a greater being. Such a creature is able to possess a power of destruction which is truly Satanic in the sense that "Satan is the ape of God"; for such a human type wields, at least for a short time, a godlike but destructive dominion over the earth. This dominion is devoid of the care which God displays toward all His creatures and bereft of that love which runs through the arteries of the universe.

As traditionally defined by Islam, humankind is seen as being given power even to the extent of finally causing corruption on earth as predicted in the Quran. But this power is seen in the traditional Islamic perspective to be limited in normal circumstances by the responsibilities which each person bears not only toward God and other men and women, but also toward the whole of creation. The Divine Law (*al-Sharīʿah*) is explicit in extending the religious duties of humans to the natural order and the environment. One must not only feed the poor but also avoid polluting running water. It is pleasing in the eyes of God not only to be kind to one's parents, but also to plant trees and treat animals gently and with kindness. Even in the realm of the Divine Law, and without turning to the metaphysical significance of nature, one can see the close nexus created by Islam between humanity and the whole natural order. Nor could it be otherwise, for the primordial character of the Islamic revelation reinstates humanity and the cosmos in a state of unity, harmony, and complementarity, reaffirming our inner bond to the whole of creation, which, as already stated, shares with us the Quranic revelation in the deepest sense.

There is so much talk today of human rights, that it is necessary to mention here the basic truth that according to the Islamic perspective responsibilities precede rights. Humans have no rights of their own independent of God, whether these rights be over nature or even over themselves, since they are not the creators of their own being. Humanity is not in fact capable of creating anything from nothing. The power of *fiat lux* belongs to God alone. What rights humans do possess are given to them by God as a consequence of their having accepted the covenant with God and fulfilled their responsibilities as God's vice-gerents on earth.

The Islamic attitude toward humanity differs profoundly not so much from traditional Jewish and Christian views, although even here there are some notable differences, as from postmedieval forms of humanism, to which much of later religious thought in the West gradually succumbed. Islam sees God alone as being absolute. One of the meanings of the testimony of Islam (*Lā ilāha illa'Llāh*) is that there is no absolute unless it is the Absolute. The human being is seen as a creature who, as a theomorphic being, reflects all of God's Names and Qualities in a direct and central fashion, but no human is absolute in himself or herself, especially in this transient earthly state. In fact, whatever positive qualities a human being possesses come from God. That is why in the Quran, it is asserted, "God is the rich and ye are the poor" (XLVII:38). Humanity's greatest glory lies in the realization of this poverty through which alone the Absolute can be reached.

In contrast, since the advent of Renaissance humanism, Western civilization has absolutized earthly man. While depriving man of a center and creating a veritable centerless culture and art, Western humanism has sought to bestow upon this centerless humanity the quality of absoluteness.[10] It is this purely earthly man defined by rationalism and humanism who developed seventeenth-century science based upon the domination and conquest of nature, who sees nature as an enemy, and who continues to plunder and destroy the natural environment always in the name of the rights of man, which are seen as absolute. It is this terrestrial man, now absolutized, who destroys vast forests in the name of immediate economic welfare without thinking for a moment of the consequences of such an action for future human generations or for other creatures of this world. It is such a creature who, in seeing earthly life as being absolute, tries to prolong it at all cost, creating a medicine which has produced both wonders and horrors including the destruction of the ecological balance through human overpopulation. Neither God nor nature have rights for a humanity which sees itself as absolute.

Now, Islam has always stood strongly opposed to this absolutization of what one might call the Promethean and Titanic man. It has never permitted the glorification of humanity at the expense of either God or His creation. Nothing is more detestable to traditional Muslim sensibilities than some of the Renaissance Titanic art created for the glorification of a humanity in rebellion against heaven. If modern science and with it a civilization which gave and still gives itself absolute right of domination over the earth and even the heavens did not come into being in the Islamic world, it was not because of the lack of mathematical or astronomical knowledge. Rather, it was because the Islamic perspective excluded the possibility of the deification of earthly man or the total secularization of nature. In Islamic eyes, only the Absolute is absolute.

The consequence of this perspective upon the relation between humankind and the environment has been immense. In the traditional Islamic world, since the human state was never absolutized, human rights were never made absolute in total forgetfulness of the rights of God and also of His other creatures. The modern Westerner, in contrast to the traditional Muslim or for that matter Christian, owes nothing to anyone or anything. Nor, as already mentioned, does he or she feel responsible to any other being beyond the human. In contrast, the traditional Muslim or *homo islamicus* has always lived in an awareness of the rights of God and of others, including the nonhuman realm, and remained conscious of having a responsibility to God and also for His creatures. Islam has been always strongly opposed to rationalism while being rational, to naturalism while being aware of the sacred quality permeating the natural order, and to humanism while being concerned with humanity and its entelechy in the deepest manner possible. These attitudes, moreover, exercised an immense influence upon the Islamic attitudes toward nature and the natural environment, especially before the domination of the Islamic world by the West.

Many secularists in the West today blame what is now called the Judeo-Christian tradition, to which Islam is also added in this context and not elsewhere, for the present ecological crisis. They forget that neither Christian Armenia, nor Ethiopia, nor even Christian Eastern Europe gave rise to the science and technology which, in the hands of secular man, has led to the devastation of the globe, and that therefore other factors must have been involved. Moreover, it must first be remembered that, if one chooses not to speak of Judaism and Christianity but of the Judeo-Christian tradition, one should speak of the Judeo-Christian-Islamic tradition, which would thus include the three members of the Abrahamic family of religions. Second, one must remember that each of the religions of the Abrahamic family has its particular doctrinal and theological emphasis and spiritual contour. As far as the question of the spiritual and metaphysical significance of nature is concerned, Islam has placed greater emphasis upon it than the mainstream theological tradition of Western Christianity and has always emphasized and preserved even to this day teachings which have been either forgotten or marginalized in religious thought in the West.

This does not mean, however, that Judaism or Christianity are in themselves responsible for the environmental crisis. Although Western Christianity did fail to emphasize the spiritual significance of nature in its mainstream theology even before the Renaissance, it also battled for five centuries against humanism, rationalism, and secularism, with their view of the cosmos and approval of the rape of nature. Western Christianity's eventual acquiescence to these schools of thought has led to the successful development of a secular science and technology which have forgotten the sacred quality of nature and lost the metaphysics which alone can explain who humans are, why their rights are limited, and why they are the bridge between heaven and earth, called to be custodians of the earth and its creatures.

It must now be asked why, if the traditional teachings of Islam concerning the natural order as outlined above are still alive, they are not more evident in voices from the Islamic world, nor more effective in the practical realm in averting ecological disasters? And why is it that the environmental crisis is no less acute in the Islamic world than in other parts of the globe? Let us first of all turn to the voices from the Islamic world which the West has heard during the past century and a half and still hears, and through which it interprets the Islamic view concerning the natural environment.

During the last century and a half, two voices from the Islamic world have been most vociferous and easily heard in the West: the voice of the so-called fundamentalist reformers and that of the modernists.[11] The first includes such schools as the Wahhabīs and Salafīs, who have stood opposed to the West and defended the sacrosanct character of the Divine Law, seeking to reestablish a society in which this law would be promulgated thoroughly and completely. At first, its proponents were against Western technology, as can be seen in the attitude of the Wahhabī-supported Saudis in Arabia during the first decades of this century. But this opposition was more juridical than intellectual. These movements were not usually concerned with the traditional Islamic philosophy of nature. They dealt with the environment according to *Sharīʿite* norms, but without the critical knowledge of Western science and technology necessary to avert the catastrophic effects of modern science upon the religiously inspired views of nature and of modern technology upon the environment. Furthermore, they were too engrossed in combating Western colonial influences and in what they considered "cleansing" Islamic society of alien accretions to be much concerned with the natural environment.

It was this lack of knowledge and critical judgment which led to an open espousal of Western science and technology by later followers of this very group during the second half of this century. Again this can be seen in Saudi Arabia, which began to embrace rapid industrialization from the 1950s onward while maintaining its close links with Wahhābism. Concern with the environment there did not in fact become an issue until very recently.

The second voice, which is that of the modernists, expressed a staunch defense of Western science and technology from the early nineteenth century. The defeat of Egypt by Napoleon in 1798, followed by the defeat of the Ottomans and the Persians by European powers and the British conquest of India in the early decades of the nineteenth century, led to a crisis within the Islamic world, which was not only political but also cultural and religious. As political leaders like Muḥammad ʿAlī of Egypt sent students to Europe to master Western military arts, modernist thinkers began not only to accept but practically to idolize Western science and technology, which they saw as the secret of the West's power. From Sir Sayyid Aḥmad Khān in India to Muḥammad ʿAbduh in Egypt, from Zia Gökalp in Turkey to Seyyed Hasan Taqizadeh in Persia, modernists stressed that Western science and technology would lead to the material and even spiritual happiness of Muslims. A figure such as Jamāl al-Dīn al-Afghānī simply equated Western with Islamic science, which is mistaken because, although Western science owes a great deal historically to Islamic science, it has never accepted the philosophical framework of Islamic science. For over a century, teachers in classrooms and even preachers from pulpits of mosques repeated this view, extolling Western science and technology and considering its mastery as practically a religious duty. There were a few dissenting voices here and there, such as Sayyid Aḥmad Kasrawī in Persia, who openly criticized Western science and technology, but they were brushed aside by the modernists as being simply obscurantists.[12]

Meanwhile, other voices, which are those of traditional Islam, especially in its sapiential, not simply juridical dimension, survived even though they were hardly heard in the West until quite recently. Poets still expressed the traditional love of nature, and those devoted to the inner dimension of Islam continued to study the cosmos as a book to be deciphered and understood by penetrating into the inner meaning of its symbols. But until recently the West hardly heard these voices. Occasionally, a poet such as Muḥammad Iqbāl would become well known in the West, but he would not be the type of poet to sing primarily of the love of nature. As for Iqbāl, he was too deeply engrossed in the current problems of the Islamic community and too profoundly influenced by nineteenth-century European philosophy to emphasize that science should lead to the contemplation of nature and spiritual perfection, rather than to dominion and power over nature. Yet, here and there in his poetry, rather than his prose, one can still gain a glimpse of that attitude toward nature which Sufis and Islamic philosophers have cultivated over the centuries on the basis of the clear message of the Quran.

The voice of traditional Islam in its sapiential dimension, wherein is to be found the Islamic doctrine of the ultimate meaning of nature, continued to resonate within the Islamic world although it was now no longer the dominating voice. Nor was it heard by the West, which until recently devoted its study of the Islamic world almost completely to the fundamentalist reformers and modernists. These two opposing groups have disagreed on many issues during the past few decades, but they have shared a blind acceptance of modern science and technology and a total neglect of traditional Islamic views concerning science and nature. As the ecological crisis has become a major global issue, however, the voice of traditional Islam has begun to be heard ever more clearly and loudly. It is this voice which speaks of the millennial wisdom of Islam and its science in treating the natural environment and which insists that the role of religion in the present environmental crisis is not only ethical but also intellectual. It thus offers an in-depth critique of the totalitarian and monopolistic contention of modern science as the only valid form of knowledge about the natural domain.

As stated already, there are also practical reasons why the Islamic world has not been more successful on the operational level in avoiding an environmental crisis, despite the religiously positive and caring attitude of Islam toward nature. These reasons are related to the global domination of the West and the need felt not only by Muslims but by what is wrongly called "Third World countries" to overcome the economic consequences of this domination. This need cuts across several continents and can be seen in Islamic Egypt and Buddhist Thailand, Hindu India and Christian Ethiopia. This worldwide need, added to common human nature, which seeks everywhere the forbidden fruit of Faustian science, has caused Shinto-Buddhist Japan with its remarkable appreciation of nature and the

Navajo Nation with its incredible spiritual insight into the significance of natural forms to suffer almost as much from the destruction of their natural environment as formerly Communist and even Catholic Poland or half-secularized and half-Christian northern New Jersey. The fact that Cairo and Karachi suffer from environmental decay does not negate traditional Islamic doctrines concerning the love and appreciation of nature any more than the pollution of Tokyo negates the spiritual significance of the Zen gardens of Kyoto.

The economic and political factors which have prevented Muslims from paying greater attention to their own traditional teachings concerning the environment are very complex and need a separate treatment. Suffice it to say that when the pollution of the Hudson River can be measured in the Azores, when every American born will use over a hundred times more of many forms of raw material than a Muslim from Bangladesh, when the West refuses to put its own house in order through a sane energy and raw materials policy and instead invades another part of the globe to preserve "the Western way of life" (which in this context means the wasteful use of energy without thought of the consequences for future generations), then it should not be too difficult to understand why the environmental issue does not possess much priority in the Islamic world at the present moment.

Furthermore, there is the question of innovation and reception of a Western technology which until now has been very destructive of the natural environment. The Islamic world is on the receiving end of an ever-changing technology which guarantees its domination through constant innovation. There is no breathing space in which to adapt an existing technology with the minimum environmental impact without suffering severe economic pressure, which is simply unbearable for most Islamic countries. Were "catching up" with Western technology, which is the goal of many Muslim governments, actually possible, such an achievement would simply expand the circle of the virulent and aggressive destruction of the environment caused by such technology, as the cases of Japan, Korea, and Taiwan bear out. Such countries have not become ecologically safer havens since joining the industrialized world. One wonders what would happen to the world's biosphere were people from Nigeria to Indonesia to spend the same amount of energy and use the same amount of raw materials as citizens of the so-called advanced countries!

On the practical level, there is finally one other important factor to mention as far as the Islamic world is concerned. Colonial domination by the West did not only bring about economic exploitation and introduce a second-rate Western technology into the Islamic world. It also resulted in many Muslim countries discarding much of the Divine Law or *al-Sharīʿah* with its numerous teachings concerning responsibility toward the natural environment in favor of secular Belgian, French, or British codes, which had little to say concerning nature. And even if laws relating to the environment were passed in the framework of the new secular laws imported from the West, these did not carry any religious weight and remained divorced from ethical considerations, whose sole origin and source for Muslims is the Islamic revelation. This change in the significance of law in many Islamic lands, added to the migration of a large number of people from the countryside to urban areas with its concomitant cultural dislocation, poses a major obstacle to the propagation of traditional Islamic teachings concerning the environment. The callousness toward domestic animals and trees evident today in many of the large Middle Eastern cities reveals the existence of these factors. The great poverty in many Islamic areas also works against reawakening an active concern for nature. The problem of the preservation of the natural environment seems simply too far removed from the immediate concerns of life for those under these influences, and many political leaders simply relegate it to a position of secondary importance and consider it to be merely a Western problem despite blatant evidence to the contrary.

Of course, the environmental crisis is not only Western but also global. And although Muslims for the most part endanger only themselves in their heedless attitude toward the environment, while the highly industrialized countries threaten the ecology of the whole globe, it is absolutely essential for the Islamic world to face this issue in the most serious manner.[13] It is also important for everyone to realize that since the environmental crisis is global, it requires global attention. The Islamic world must do its utmost to bring its rich intellectual and ethical tradition to bear upon this problem as the West must realize that there is a wisdom within the Islamic tradition concerning nature and the natural environment which can be of great significance for those in other traditions reformulating a theology of nature.

At this point, it is necessary to point to some of the elements which distinguish the situation of Islam today from that of the West as far as the environmental crisis is concerned. In the West, there is a need to reformulate a Christian theology of nature in a society in which for several centuries religious faith has weakened and theology has surrendered the realm of nature to science and shied away from any serious concern with the sacred dimension of natural forms and phenomena. There is also the need to dethrone the humanistic conception of man, which makes of man almost a deity who determines the value and norm of things, and who looks upon all of nature from only the point of view of self-interest. This dethronement means a death of the type of person who almost instinctively views nature as the enemy to be conquered and the birth of the person who respects and loves nature. This new respect and love will provide people spiritual as well as physical sustenance from nature, while also enabling them to give something of themselves to the multifarious species of the natural kingdom. As far as the environmental crisis is concerned, any change short of this death and rebirth of modern Western man is cosmetic. No amount of clever engineering based on the current secular science of the natural order will be able to avert the catastrophe created by the applications of this science.

Seeing how deeply embedded alienation from nature, as well as from the supernatural source of the natural order, is in the secularized Western psyche and mind, the task of teaching the Western world to love and respect nature is an extremely difficult one. But there is a compensation in the fact that the forces with which the religious and spiritual elements in the West have to contend come for the most part from within Western civilization itself and not from the outside, the economic and now technological challenge of Japan being the only exception.

In contrast, in the Islamic world the origins of technological problems bearing upon the environment are to be sought outside that world. The intellectual and spiritual leaders of the Islamic world must deal not only with their own problems but with ever continuing challenges which originate beyond their borders. There are, however, also certain advantages in the Islamic situation. There, religion is still very strong, and one can appeal more easily to people's religious sensibilities in seeking to solve the environmental crisis. Moreover, what would be called a theology of nature in Christian terms has never been forgotten in Islam; nor has the sacred view of the cosmos become totally replaced by a view based upon a purely quantitative science as has occurred in the West. Finally, the ethical dimension of life as grounded in revelation is still strong among Muslims and can be appealed to more easily than is the case in many but, of course, not all sectors of Western society. The task of saving the natural order from a humanity which has lost its vision of who it is and has thus become useless from a spiritual point of view is daunting in both worlds. But it is a task which must be carried out if human life is even to continue, let alone whether we are to lead lives with a qualitative dimension.

In conclusion, it is necessary to mention what it is exactly that the Islamic world must do in the face of the devastating environmental crisis, leaving the agenda of the West to others who are specifically concerned with and speak for it. The Islamic world must carry out two extensive programs despite all the obstacles placed before it by external factors.[14] The first concerns formulating and making clearly known in a contemporary language the perennial wisdom of Islam concerning the natural order, its religious significance, and intimate relation to every phase of human life in this world. This program must of necessity include a critical appraisal of both modern science and scientism, as well as an examination of the significance of traditional Islamic science, not only as part of the Western history of science, but also as an integral part of the Islamic intellectual tradition.[15]

The second program needed is expansion of the awareness of *Sharīʿite* teachings concerning the ethical treatment of the natural environment and extension of their field of application whenever necessary according to the principle of the *Sharīʿah* itself. In addition to passing laws of a civil nature against acts of pollution, similar to what is done in the West, Islamic countries must bring out the *Sharīʿite* injunctions concerning care for nature and compassion toward animals and plants, so that environmental laws will be seen by Muslims to be impregnated with religious significance. The ethical treatment of the environment in the Islamic world cannot take place without an emphasis upon the teachings of the Divine Law, and hence upon the ethical and religious consequences for the human soul if the natural environment is treated with impunity and raped with unbridled ferocity.

In traditional times, there existed not one but several humanities, each dominated by a religious and spiritual norm which could be called "the presiding Idea" of the civilization in question. The religion of one culture remained impervious to other universes of religious discourse, with exceptions which only proved the rule.[16] Today the boundaries of those traditional universes have been broken, and there is a need for them to understand each other and to reach a harmony which is in fact possible, to paraphrase Frithjof Schuon, not in the human atmosphere but in the Divine stratosphere. Meanwhile, however, the members of these different human collectivities have nearly all become participants, some more actively than others, in the destruction of the earth. It is, therefore, essential for those who speak for religion and the world of the spirit to collaborate and apply that inner unity and harmony which binds them together to the terrestrial realm and to the question of saving the planet from a humanity in rebellion against both heaven and earth. The person who speaks for the life of the Spirit today cannot remain indifferent to the destruction of that primordial cathedral which is virgin nature, nor maintain silence concerning the harm human beings do to themselves as immortal beings by absolutizing the "kingdom of man" and as a consequence brutalizing and destroying everything else in the name of the earthly welfare of members of that kingdom.

Islam certainly has its share of responsibility in drawing the attention of its own adherents, as well as the world at large, to the spiritual significance of nature and the necessity to live in peace and harmony with the rest of God's creation. The Islamic tradition is particularly rich in preserving to this day a sapiential knowledge combined with love of the natural environment, a metaphysics of nature which unveils her role as the grand book in which the symbols of the world of Divine majesty and beauty are engraved.

It also possesses an ethics, rooted in revelation and bound to the Divine Law, which concerns the responsibilities and duties of humanity toward the nonhuman realms of the created order. It is incumbent upon Muslims to resuscitate both of these dimensions of their tradition in a contemporary language which can awaken and lead men and women to a greater awareness of the spiritual significance of the natural world and the dire consequences of its destruction.

It is also the duty of those who speak for traditional Islam to carry out a dialogue with followers of other religions on an issue which concerns men and women everywhere. By sharing the wisdom of their tradition with others and learning from them, Muslims can contribute a great deal not only to the Islamic world itself as it struggles with the consequences of the environmental crisis, but also to the whole of humanity. As the sun shines upon all men and women from East to West and the night stars reveal their mysterious beauty to those with eyes to see, whether they behold them in Japan, India, Arabia, or America, so does the wisdom concerning nature and the compassionate care for nature as taught by various religions belong to human beings wherever they might be, as long as they are blessed with the gift of appreciation of the beauty of the rose and the song of the nightingale. The Quran asserts that "to God belong the East and the West" (II:115). This statement possesses many levels of meaning, one of which is that where the sun rises and where it sets, where forests cover the land and where sand dunes rove over empty spaces, where majestic mountains touch the void of heaven and where deep blue waters reflect the Divine Infinitude, all belong to God and are hence interrelated. The destruction of one part of creation affects other parts in ways that the science of today has not been able to fathom. In such an interdependent natural environment, in which all human beings live, it is for men and women everywhere to unite, not in an agnostic humanism which kills the Divine in man and therefore veils the reflection of the Divine in nature, but in the one Spirit which manifests itself in different ways in the vast and complex ocean of humanity.

To rediscover this Spirit and its reflection in oneself is the first essential step. To see the reflection of this Spirit in the world of nature is its natural consequence. Humanity cannot save the natural environment except by rediscovering the nexus between the Spirit and nature and becoming once again aware of the sacred quality of the works of the Supreme Artisan. And men and women cannot gain such an awareness of the sacred aspect of nature without discovering the sacred within themselves and ultimately the Sacred as such. The solution of the environmental crisis can come about only when the modern spiritual malaise is cured and there is a rediscovery of the world of the Spirit, which, being compassionate, always gives of Itself to those open and receptive to Its vivifying rays. The bounties of nature and its generosity to the human race are there as proofs of this reality, for despite all that we have done to destroy nature, she is still alive and reflects on her own ontological level the love and compassion, the wisdom and the power, which belong ultimately to the realm of the Spirit. And in this crisis of unprecedented proportions, it is nature as God's primordial creation that will have the final say.

NOTES

1. Concerning the Islamic view of nature see Seyyed Hossein Nasr, *An Introduction to Islamic Cosmological Doctrines* (London: Thames and Hudson, 1978); *Science and Civilization in Islam* (Cambridge: Islamic Text Society, 1987); and *Islamic Life and Thought* (Albany: SUNY Press, 1981), especially chap. 19.
2. On the concept of Islam as the primordial religion see Frithjof Schuon, *Understanding Islam*, trans. D. M. Matheson (London: George Allen & Unwin, 1979); and Seyyed Hossein Nasr, *Ideals and Realities of Islam* (New York: Harper Collins, 1989). Islam is also called *dīn al-fiṭrah* which means the religion that is in the very nature of things and engraved in man's primordial and eternal substance.
3. See my "The Cosmos and the Natural Order," in Seyyed Hossein Nasr, ed., *Islamic Spirituality—Foundations*, vol. 19 of *World Spirituality—An Encyclopedic History of the Religious Quest* (New York: Crossroad, 1987), 345ff. See also Nasr, *Ideals and Realities of Islam*, 53ff.
4. See Nasr, *Introduction to Islamic Cosmological Doctrine*, 6ff, and also Nasr, *Ideals and Realities of Islam*, 55.
5. See William C. Chittick, "'God Surrounds all Things': An Islamic Perspective on the Environment," *The World and I*, 1, no. 6 (June 1986): 671–78.
6. It is of great significance that the Islamic paradise is not constructed only of precious stones from the mineral realm, but also contains plants and animals.

Certain later Islamic philosophers, such as Ṣadr al-Dīn Shīrāzī in his *al-Asfār al-arbaʿah* and *Risālah fi 'l-ḥashr*, speak at length of the resurrection of animals as well as of people.

7. Trans. Annemarie Schimmel, in her *As Through a Veil: Mystical Poetry in Islam* (New York: Columbia University Press, 1982), 150.
8. On the Sufi attitude toward nature, see F. Meier, "The Problems of Nature in the Esoteric Monism of Islam," in *Spirit and Nature: Papers from the Eranos Yearbooks*, trans. R. Mannheim (Princeton: Princeton University Press, 1954), 203ff; and Nasr, *Islamic Life and Thought*, chap. 19.
9. On the Islamic concept of man, see G. Eaton, "Man," in Nasr, ed., *Islamic Spirituality—Foundations*, chap. 19; also Seyyed Hossein Nasr, *Knowledge and the Sacred* (Albany: SUNY Press, 1989), 358–77.
10. For a profound study of this loss of center in the West as a result of the advent of humanism, see Frithjof Schuon, *To Have a Center* (Bloomington, Ind.: World Wisdom Books, 1990), 160ff.
11. This categorization is somewhat simplified for the sake of argument. There are, needless to say, shades of opinion and a certain amount of diversity in each category although the general characteristics of each voice as outlined below hold for members of the category in question.
12. Kasrawī, who died in 1946, held many views which are problematic from a traditional Islamic point of view, but he was perhaps the first Muslim writer to have criticized thoroughly European science and technology and their effect upon society.
13. I say "for the most part" since most of the environmental problems of the Islamic world concern the health and physical well-being of its own citizens and do not affect the pollution of the global environment in the same way and to the same degree as do the actions of highly industrialized nations. There are, however, actions taken by certain Islamic, or for that matter so-called Third World, countries which do have dire consequences for the globe as a whole. These include the destruction of tropical forests and the use of pesticides, which are of course the products of modern chemistry and part and parcel of modern agricultural practice. Strangely enough, because of such practices and their global consequence, for the first time since the beginning of the period of European expansion, the life of the industrialized West depends in a basic way upon the actions of those who live in what used to be European colonies.
14. I would be the last to advise inaction by Muslims because of the technological and economic domination of the West. Even under the difficult conditions of today, much can be done by Muslims themselves. Moreover, some of the ecologically catastrophic actions taken in various Islamic countries, ranging from

agricultural to architectural fiascos, did not have to be carried out. Despite the West's grave responsibility as the main agent in the creation of the present global environmental crisis, it cannot be blamed for every instance of mistaken planning and action, nor for inaction, within Islamic countries.

15. See my *Islamic Life and Thought* and *Science and Civilization in Islam* for an extensive treatment of this subject.
16. See Frithjof Schuon, *The Transcendent Unity of Religions*, trans. P. Townsend (Wheaton, Ill.: Theosophical Pub. House, 1984), especially chaps. 2 and 3.

CHAPTER 6

A Tibetan Buddhist Perspective on Spirit in Nature

Tenzin Gyatso, His Holiness the 14th Dalai Lama

Fig. 6. This Tibetan Buddhist appliqué, Vajrabhairava *(ca. 1800), which was designed for Tantric meditation practices, employs powerful physical imagery to symbolize the true self (the essence of buddha-nature) and the most profound form of spiritual liberation. It depicts the Bodhisattva of Wisdom, Mañjuśrī, in the wrathful form of Vajrabhairava, the Indestructible Terror who triumphs over evil, suffering, and death. His appearance as an enraged bull is intended to frighten away egocentric delusions. The sexual union of Vajrabhairava with his female consort, Vajravetali, symbolizes the bliss of full enlightenment as involving the intimate union of compassion and wisdom. (Collection of Mr. and Mrs. John Gilmore Ford, Baltimore, Maryland.)*

The first of the basic Buddhist precepts counsels those pursuing the path to liberation to avoid destroying life, and one of the most popular Buddhist scriptures, the Mettāsutta, urges the faithful to "develop loving-kindness for the entire world." These ethical teachings, reinforced by a worldview that emphasizes the interdependence of all beings and the immanence of the sacred, have generated much interest in Buddhism in this time of environmental crisis. The most widely known spokesperson for the Buddhist tradition in the world today is the Dalai Lama, whose teachings strive to develop in others a sense of universal responsibility, emphasizing a commitment to world peace, human rights, and protection of the environment.

Tenzin Gyatso was born into a Tibetan peasant family home, but at the age of two he was identified, in accordance with traditional Tibetan practices, as the reincarnation of his predecessor, the thirteenth Dalai Lama. Soon thereafter he entered a rigorous program of education and spiritual training, which led him to earn the Geshe degree, a doctorate in Buddhist philosophy. At the age of fifteen, before his formal education was complete, he was called to assume full political responsibility as head of the Tibetan government. He found his homeland facing a growing threat from China. When the situation worsened and the Chinese army occupied Tibet in 1959, he was forced into exile. Since then he has resided in Dharmsala, India, where he heads the Tibetan government-in-exile and presides over a large community of Tibetans.

In an effort to resolve differences with the Chinese government and to restore human rights and peace in Tibet, the Dalai Lama has proposed a Five Point Peace Plan. The central idea in his proposal is the recommendation that Tibet be transformed into a zone of ahimsa (nonviolence), a peace sanctuary. More specifically this would mean withdrawal of all military forces from Tibet, establishment of democratic self-government, and the preservation of Tibet as a great natural park and wilderness sanctuary where plant and animal life are strictly protected. "It is my dream," writes the Dalai Lama, "that the entire Tibetan plateau should become a free refuge where humanity and nature can live in peace and in harmonious balance." He explains that the transformation of Tibet into a peace sanctuary in Asia would fulfill its historical role as a peaceful Buddhist nation and create a needed buffer region separating the rival powers of India and China.

In his essay the Dalai Lama describes the earth as our home and seeks to set forth a "practical ethic of caring for our home" grounded in the Buddhist understanding of interdependence. It is his view that the most important factor in developing an ethic of caring for the earth is the cultivation in individuals of an attitude of compassion. Compassion generates the sense of universal responsibility that is fundamental to the true nature of humanity. With the growth of this sense of responsibility, there also arises the hope, courage, joy, and inner peace needed to sustain a person engaged in the problems of the contemporary world.

BROTHERS and sisters, I am very happy to be here with you, to come to this beautiful place once more. I have enjoyed the last few days very much. The speeches from the leaders of the various traditions have been very impressive.

Although I have prepared a speech myself, a large number of people have come here today, and I think you may have different interests. This creates confusion in my mind right now; just what subject should I address to be most helpful to all of you?

The first thing that will relieve my small anxiety is to confess that I am not an expert or a specialist on ecology or the environment. So I will address a broader subject. And if you have come here with some expectation on that score, I can say that, essentially, I have nothing to offer to you. I can simply try to share some of my own views and experiences, and then maybe some of you will find some benefit; or at least some new ideas to think about.

Now, first I will try to explain briefly the Buddhist attitude and approach to the environmental crisis. In dealing with this subject I would like to divide my talk in three stages. First, I will talk about the Buddhist perception of nature and reality. Second, I will discuss what kind of ethical principle an individual should adopt, based on that view of reality and nature. Third, I will talk about what kind of right conduct, what kind of measures, individuals and society should take to restore and correct the degradation of nature and the earth, based on such an ethical principle.

When talking about developing a correct understanding or correct view of reality and nature, Buddhism emphasizes the application of reasoning and analysis. It talks about four avenues of reasoning or analysis through which one can develop a correct understanding of reality and nature. These four can be called natural, relational, functional, and logical avenues of reasoning. Reasoning and analysis have to take into account the natural laws of the universe, the interrelationships that exist in the universe, the functional properties of things in reality, and the processes of reason itself, with which it understands the universe.

First one takes into account the fundamental laws of nature, such as the fact that things exist, the fact that matter differs from consciousness, the fact that mind exists in a certain way, and so on. Second, reason takes into account the interdependence between these various entities that exist in the world, the interdependence between causes and conditions, the interdependence between parts that constitute a whole, and so on. Third, reason takes into account the functional properties that we see in reality, the properties which emerge as a consequence of the interaction between multiple factors. Fourth, based on these three levels of understanding nature, Buddhism emphasizes understanding the process of human reasoning and analysis itself. For example, reason can understand how reliable knowledge is generated through inference, either about the probable nature of a cause from the observed nature of its effect, or about a probable future state of affairs from an observed state of its cause. In short, while Buddhism is usually thought of as a religion, it is actually a way of thought that emphasizes the necessity for human reason to be applied to human problems.

When talking about the fundamental nature of reality, one could sum up the entire understanding of that nature in a simple verse: "Form is emptiness, and emptiness is form" (*The Heart Sūtra*). This simple line sums up the Buddhist understanding of the fundamental nature of reality.

In appearance, we see the world of existence and experience. In essence, all those things are empty of intrinsic reality, of independent existence.

Superficially, if we were to look at the words "emptiness" and "form" or "appearance," they might seem to be contradictory. If anything has appearance, how can it be empty? If anything is empty, how can it have a form or appearance? To overcome this contradiction, one must understand the meaning of emptiness to be interdependence. The meaning of interdependence is emptiness of independent existence. Precisely because things and events exist relatively and appear as having form, they are empty of independent existence.

Events and things come into being as a result of the aggregation of many factors—causes and conditions. But because they lack independent or absolute existence, it is possible for experiences such as our sufferings—which we do not desire—to come to a cessation. And because they lack independent or absolute existence, it is possible for pleasant experiences such as our happiness—which we do desire—to be created within ourselves.

Fundamental to attaining the Buddhist perception of reality, which ultimately is emptiness, is the understanding of relativity, the principle of interdependence. And the meaning of interdependence has three levels. At its subtlest level, it is the interdependence of things with thought and conceptual designations. At its middle level, it is the interdependence of parts and wholes. And at the surface level, it is the interdependence of causes and effects.

I think there is a direct connection between the correct understanding of ecology and the natural environment and the Buddhist principle of interdependence in terms of causes and effects and in terms of parts and wholes, factors and aggregates. But the correct understanding of the subtlest level of interdependence—that of the interdependence of things and conceptual constructions—has more to do with maintaining the balance of the outer and the inner world, and with the purification of the inner world.

I believe that every individual living being, whether animal or human, has an innate sense of self. Stemming from that innate sense of self, there is an innate desire to enjoy happiness and overcome suffering. And this is something which is innate to all beings. I believe it is a natural phenomenon. But if we tried to examine why such innate faculties are there within living beings, I do not think we could ever find a convincing answer. I would rather stop there and say that it is a natural fact. Various different philosophies have tried to examine that nature of living beings. And still, after centuries, this is not yet finally solved. So I think it is better to accept this as something natural, as a reality.

Therefore, we can say that the purpose of life is happiness, joy, and satisfaction, because life itself, I think, exists on the ground of hope, on the basis of hope. And hope is, of course, for the better, for the happier. That is quite natural, isn't it? In that case, relations with one's fellow human beings—and also, animals, including insects (even those which sometimes seem quite troublesome)—should be based on the awareness that all of them seek happiness, and none of them want suffering. All have a right to happiness, a right to freedom from suffering.

And generally speaking, all beings seem beautiful to us, beautiful birds, beautiful beasts. Their presence gives us some kind of tranquillity, some kind of joy; they are like an ornament to our lives really. And then the forest, the plants, and the trees, all these natural things come together to make our surroundings pleasant. All are heavily interdependent in creating our joy and happiness, in removing our sufferings.

Our human ancestors survived by depending on trees, on wood. Their fires depended on the wood. The trees gave them shelter and protection. When a dangerous animal threatened them, they could climb up to safety. Some trees bear beautiful flowers, which are ornaments, which they picked and wore in their hair, something like our modern jewelry. Then, of course, there's the fruit of the trees, and nuts, which are nourishing. And finally, of course, there are sticks made from the branches; when someone attacks, it's a weapon; when you get older, it is a cane, like a reliable friend.

Such examples, I think, show the historical basis of human nature. Later, as human culture developed, we made something more beautiful out of it, something poetical. During our ancestors' time, human survival and welfare were very dependent on trees. But as society and culture became more developed and sophisticated, this dependence became less and less, and trees became the subject of poetry.

So, therefore, this shows that our very existence is something heavily dependent on the environment.

Now since we are seeking happiness and joy, we must be able to distinguish the different causes and conditions that lead to happiness and joy, causes both immediate and long-term. One finds that, although the ultimate aim of the major world religions is the achievement of the happy life after death, eternal life, they do not encourage their adherents to neglect the well-being of the present life.

The expressed aim of Buddhism is the purification and development of the mind through mental training in order to attain supreme liberation. But the meditation manuals place great emphasis on finding an ideal environment for the practice of training the mind because a cleaner environment does have a tremendous impact on one's spiritual progress. The Buddhist literature mentions the sanctity of the environment as inspiring and blessing the practitioner, and in turn the practitioner's spiritual realization blessing the environment. There is an exchange between human spirit and nature. In tune with such awareness, we find in Buddhist practice specific rituals aimed at regenerating the vitality of the earth, at purifying the environment, wherein certain precious minerals are buried underground, and then consecration rituals are performed.

I think that in ancient time, the human ability to measure the imbalance of nature was very limited—almost none. At that time, there was no need for worry or concern. But today, the human ability to disturb the balance of nature is growing. World population has increased immeasurably. Due to many factors, nature, even the Mother Planet herself, it seems, is showing us a red light. She is saying, "Be careful, you should realize there are limits!"

Taking care of the planet is nothing special, nothing sacred or holy. It's just like taking care of our own house. We have no other planet, no other house, except this one. Even if there are a lot of disturbances and problems, it is our only alternative. We cannot go to any other planet. If the moon is seen from a distance, it appears quite beautiful. But if we go there to stay, I think, it would be horrible. So, our blue planet is much better, much happier. Therefore, we have to take care of our own place. This is not something special or holy. This is just a practical fact!

Now I will go on to the second part of my talk, the development of an ethical principle based on the Buddhist understanding of reality and nature as emptiness and interdependence.

Essentially, nature's elements have secret ways of adapting. When something is damaged, another element helps out and improves the situation through some kind of evolution. This is nature's way of adjustment. But then, human intervention creates certain changes which do not give nature and its elements time to cope. So the main troublemaker, the major cause of imbalance, is we human beings ourselves. Therefore, the responsibility should be borne by us. We must find some way to restrain our destructive habits.

We cause these problems mainly with our modern economy. With different kinds of factories and chemicals, we have a strong negative impact on the balance of nature. The next question is, if that is the case, whether we have to stop all factories, all chemicals. Of course, we cannot do that. While there are negative side effects, there are also tremendous benefits. True science and technology bring humanity a lot of benefit.

So what to do? We must use our human intelligence. And in some cases, we must have more patience. We must cultivate more contentment. And we must handle new progress and development in a proper way, keeping the side effects to a minimum. At the same time, we must take care of the earth and its basic elements in a more balanced way, no matter how expensive the cost. I think that's the only way.

Here I have come to the third part of my talk. Based on a practical ethic of caring for our home, grounded in our understanding of interdependence, what kind of measures can we take to correct these imbalances in nature?

Generally speaking, crises emerge as a consequence of certain causes or conditions. Principal among them is ignorance of the real situation. In order to overcome that, the most effective means is to develop knowledge and understanding.

Presently, older people like myself are speaking out about these dangers—but I think that is very limited in effect. The greater responsibility, I feel, lies with the scientists, especially those who are trained in this field. Through their research, with their experimental data, they should make clear the real long-term consequences of certain negative practices and positive measures. Scientists and environmental experts should prepare a very specific and detailed global study of the long-term dangers and benefits our society will face in the future.

Materials based on such studies should then be thoroughly learned by young students in school right from the start. Young children should take the environment into account when they study about geography, economics, or history. I feel it's very important to introduce ecology into the school curriculum, pointing out the environmental problems that the world currently faces. Even at a very early age, children should be exposed to the understanding and knowledge of the planetary environmental crisis. The various media—newspapers, television—all should be responsible for communicating the reality of this threatening situation.

In some cases, we might be able to overcome ignorance, understand reality, and reach the situation where everyone knows what is going on. But still we do not act to prevent disaster. Such a lack of will to act—in spite of having the knowledge and understanding—stems, I think, either from negligence (becoming totally oblivious to the crisis) or from discouragement (the feeling that "I have no ability, I simply cannot do anything").

I firmly believe that the most important factor is our attitude and human motivation. Genuine human love, human kindness, and human affection. This is the key thing. That will help us to develop human determination also. Genuine love or compassion is not a feeling of lofty pity, sympathy tinged with contempt toward the other, a looking down on them; it is not like that. True love or compassion is actually a special sense of responsibility. A strong sense of care and concern for the happiness of the other, that is genuine love. Such true love automatically becomes a sense of responsibility.

So, how should we develop compassion? How should we expand our love? First, it is very important to know that within the meaning of "love" there are various emotions. What is commonly called "love" is often merely blind love, or blind attachment. In many cases, it involves unconscious projections on the other, possessiveness, and desire; it is usually not at all good. There is a second level of love or compassion, which is a kind of condescending pity. But that is not really positive compassion. We feel genuine compassion and love not just for beings close to us, but for all persons and animals. Such true compassion develops from the recognition that everyone does not want suffering and does want happiness, just like us. When we really feel that, we feel that they have every right to be happy and every right to overcome suffering. Realizing that, we naturally develop a genuine concern for their suffering and their right to be free from it.

We can feel this kind of genuine love for others no matter what their attitude toward us. That love is steady; so long as any person or being suffers, we feel responsible, even if he or she is our enemy. Love mixed with attachment makes us concerned only for beings close to us. That kind of love is biased and always narrow and limited. But genuine love is much wider and stronger. And it can be developed.

If we analyze the situation in various ways, we can develop a firm conviction about the need for such a mental attitude, even out of self-interest. In our daily life, it is the energy of genuine love and compassion that is the source of hope, the source of happiness, the source of joy, and the source of inner strength.

When we have that kind of love with its strong sense of responsibility, we will never lose our hope or our determination. The more we are challenged by negative forces, the more determination we will develop. So it is really the source of every success. That is what I always feel.

In our daily lives, we love smiles. I especially love a genuine smile, not a sarcastic smile, or a diplomatic smile, which sometimes even increases suspicion. But I consider the genuine smile something really precious. It is the great bridge of communication. Whether you know the same language or not, whether you are from the same culture, or nation, or race—all that is secondary. The basic thing is to realize that the other is a human being, a gentle human being who wants happiness and does not want suffering, just like ourselves. At that basic level, we just smile—we can exchange smiles. Then immediately the barrier is broken and we feel close.

After all, a human being is a social animal. I often tell my friends that there is no need to study philosophy or other complicated subjects. Just look at those innocent insects, like ants or bees. I am very fond of honey—so I am always exploiting the bees' hard work. Therefore, I have a special interest in the lives of bees. I learned many thing about them and developed a special relationship with them. They amaze me. They have no religion, no constitution, and no police force, but their natural law of existence requires harmony, and they have a natural sense of responsibility. They follow nature's system.

So what is wrong with us, we human beings? We have such a great intelligence, our human intelligence, our human wisdom. But I think we often use our human intelligence in the wrong way, we turn it in the wrong direction. As a result, in a way we are doing certain actions which are essentially contrary to our basic human nature. And here I always feel that basic human nature is compassion or affection.

This is quite simple. If we look closely at the beginning of human life, at the conception of a child, we see that sexual relations and the forming of a family are connected with real love. From the biological perspective, according to natural law, the main purpose is reproduction. And I think that the beneficial kind of love—even of sexual love—is love with a sense of care and responsibility. Mad love is not lasting, I think, if it lacks a sense of responsibility.

Look at those beautiful wild birds. When two birds come together, it is to build a nest and raise their young. When they have chicks, the male and the female both assume the same responsibility to feed the little ones. Sometimes mad love is just wild, just like dogs, completely careless about the consequences. I think it is not very good for people. If that was all there was to it, there would be no use for marriage. And yet look how people consider the marriage ceremony something very important. If we really consider it important, then we should have the love that is a sense of responsibility. If we did develop that, I think there would be fewer divorces, wouldn't there? Marriages would last longer—I think until death.

At any rate, we can see that human life begins with affection, with love, a sense of responsibility and care. We are in the mother's womb for many months. During this time the mother's mental calmness is said to be a very important factor for the healthy development of the unborn child. And after birth, according to some neurobiologists, the first few weeks are the most important period for the healthy development of the child's brain. And they say that, during that time, the mother's actual physical touch is a crucial factor. This does not come from religious scripture or ideology. It is from scientific observation.

Therefore, I believe that this human body itself very much appreciates affection. The first action of the child is the sucking of the mother's milk. And the mother, in spite of pain or exhaustion, is very willing to give milk to her child. So milk is a profound symbol of affection. Without mother's milk we cannot survive. That is human nature.

During the process of education, it is quite easy to notice how much better we learn from a teacher who not only teaches us but also shows a real concern for our welfare, who cares about our future. The lessons of such a teacher go much deeper in our mind than lessons received from a teacher who just explains about the subject without any human affection. This again shows the power of affection in nature.

The art of medicine is another good example. During this trip, I visited a hospital in New York about a problem in my left nostril. The doctor who examined me and removed the blockage was so gentle and careful, in addition to having a beautiful machine. His face was full of life—and he had a genuine smile. In spite of some pain from that small operation, I felt very fresh, quite happy and confident. In some cases when we visit doctors, they may be very professional, but if they show no human affection, we feel anxious, suspicious, and unsure how it will turn out. Haven't we all noticed that?

In our old age we again reach a stage where we come to depend heavily on others' affection. We appreciate even the slightest affection and concern. And even when we face death, on our last day, even though all efforts are now exhausted, though there is no hope, still, if some genuine friend is there at our bedside, we feel much happier. Although there is no more time to do anything, we still feel much happier—because of human nature.

So, from the beginning of human life to the end of human life, during all those years, it is clear that human affection is the key for human happiness, human survival, and human success. What do you think? This is how I feel.

Therefore, affection, love, and compassion—they are not a matter of religion. Various religions do teach us the importance of love and compassion because the basic aim of religions is the support and benefit of human beings. Since human nature is love, since genuine love and compassion are so important for life, every religion, in spite of different philosophies, traditions, and ideologies, teaches us about love and compassion. But human affection as essential for human nature is something deeper than matters of religious belief or institutional affairs. It is even more basic for human survival and success than any particular religion.

Therefore, I always used to tell people that whether they are believers or nonbelievers, that's up to them. From a certain point of view, religion is a little bit of a luxury. If you have religion, that's very good. But even without religion, you can survive, you can manage to live and even sometimes succeed. But not without human affection; without love, we cannot survive. Therefore, affection, love, and compassion, they are the deepest aspect of human nature.

Some of you here may doubt this. You may feel that anger and hatred are also part of human nature. Yes, of course anger is a human habit. But if we carefully investigate, I think we will find the dominant force of the human mind is affection.

As I mentioned earlier, when we are first born, if the mother feels the agitation of resentment or anger toward the child, then her milk may not flow freely. I noticed when I visited Ladakh that sometimes when people milk their cow, the cow's calf is brought in the front of the cow first. This way they cheat the cow; in her mind, she is giving milk to her own baby. So that shows that there is a natural condition where without a tender loving feeling of closeness, the milk may not come. So milk is the result of affection and is blocked by anger.

Again I have another reason, if we look carefully at daily life. When something happens which horrifies our minds, a murder case or terrorist attack, it is immediately reported in all the news because an event like this makes such a forceful impression in the mind. And yet every day thousands and millions of undernourished children are given food; they are nourished and they survive another day. But no one reports that because it is something normal; it should be a routine happening. We take it for granted. These facts also demonstrate our human nature and that affection is something normal. Killing and other actions born of anger and hatred are unusual for us. And so such unfortunate events strike our minds more forcefully. The basic human nature is gentle. And so I feel that there is a real possibility to promote and develop human affection on the global level. It is not unrealistic, because it is the most important part of human nature.

Each of us is an individual, naturally a part of humanity. So human effort must begin with our individual initiatives. Each of us should have a strong sense of the responsibility to create our own small part of a positive atmosphere. At the same time, we have more powerful social methods today with which to channel individual human insight and inspiration and thus to have a wider impact. There are different organizations on the national and international levels, governments, and United Nations organizations. These are powerful channels through which to implement new insights, to mobilize new inspirations.

This kind of conference is very helpful to such an end, though it would be unrealistic to expect that a few conferences could achieve any sort of complete solution. That's expecting too much. But, the constant effort of deep thought and broad discussion is very useful and worthwhile.

CHAPTER 7

"Caring for the World"

Robert Prescott-Allen

Fig. 7. The scale and precision of Bradford Washburn's (1910–) Climbers on East Ridge of the Doldenhorn, Switzerland *(1964) implies dramatic continuities. A fine stitchery of bootprints left by the climbers parallels the ridge and echoes the snow's grain, the strata of ice. In addition, the vision of tiny human figures in the mountain vastness recalls the tradition of Chinese landscape painting, in which people are presented in nature, but without ever dominating it. (© 1964 Bradford Washburn, courtesy of Palm Press, Inc.)*

For the past two decades the United Nations and the International Union for Conservation of Nature and Natural Resources (IUCN) have engaged government leaders, private environmental groups, and scientists in a series of conferences and reports that have aimed at formulating a global strategy for halting the destruction of the biosphere and achieving a sustainable way of life. Robert Prescott-Allen's essay focuses on what is the most important new contribution to the international debate on these matters, a revised version of the World Conservation Strategy which has been prepared by IUCN, UNEP, and the World Wide Fund for Nature. A draft of this document, entitled Caring for the World, *clearly outlines the basic steps which must be taken by governments, corporations, private organizations, local communities, and individual citizens if human civilization is to achieve the goal of sustainable development. (A revised and final version of this report is being published under the title* Caring for the Earth.*)*

As both a writer and the senior consultant for Caring for the Earth, *Prescott-Allen is able to provide a concise overview of the strategy and its rationale. He focuses on critical biological, economic, political, and social policy issues in a way that complements the religious emphases of other participants at the "Spirit and Nature" symposium. Especially significant is his emphasis on the need for a "world ethic of sustainability." Because this concept is of fundamental importance for reconstructing the relationship between humanity and nature, a summary of the ethical principles proposed for inclusion in the World Conservation Strategy by Prescott-Allen and by the IUCN's Ethics Working Group, chaired by J. Ronald Engel, is given below:*

PRINCIPLES OF A WORLD ETHIC OF SUSTAINABILITY

1. The ethical principles of sustainability affirm those bonds among all people and between people and the earth that protect the community of life and the rights of individuals. These principles are based on the recognition that people are an interdependent part of nature and the larger community of life and that ecological stability and the achievement of social justice are interconnected. While these principles emphasize interdependence and community, they respect biological and cultural diversity.

2. All members of the human family have the same fundamental rights, including the right: to life and security of person; to freedom of thought, inquiry, expression, conscience, religion, assembly, and association; to an education that empowers them to exercise responsibility for their own well-being and for life on earth; to an opportunity for a sustaining livelihood, including access to the resources needed for a decent standard of living within the limits of the earth; to political enfranchisement, making possible participation in government decisions that directly affect their welfare.

3. The earth should be respected at all times, which means: to approach nature with awareness, gratitude, humility, compassion, and care; to protect its essential ecological processes and life support systems; to be frugal and efficient in resource use; to conserve bio-diversity; to be guided by the best available knowledge, both traditional and scientific; and to work co-operatively to build local communities and a world society governed by policies that ensure sustainability.

4. Every life form possesses intrinsic value and warrants respect independently of its worth to people. Human development should not be at the expense of the survival of other species. People should safeguard the habitats of endangered species. They should treat all creatures decently and protect them from cruelty, avoidable suffering, and unnecessary killing.

5. The resources of the earth and the costs of fundamental development should be generously shared, especially among regions that are poor and those that are affluent. Development of one society should not be at the expense of other societies or the integrity of nature.

6. The protection of human rights and the rights of nature are a worldwide responsibility of each person and all societies, transcending all geographical, cultural, and ideological boundaries.

7. Each generation should conserve and expand the heritage of values that it has received from the earth and human culture so that future generations may receive this heritage more securely and widely shared than before. Each generation should leave to the future a world that is freer, more just, and as rich in renewable resources as the one it inherited.

8. In the face of moral dilemmas, further development of these principles for living sustainably should be pursued under the guidance of careful experimental inquiry and a spirit of compassion.

IN a wink of time we human beings have grown from being just regular critters, lost in the evolutionary crowd, to being monsters of the universe. In monster movies cute, creepy crawlies like insects and spiders are blown up to screen-filling proportions to terrorize the inhabitants of quaint little towns like Middlebury, Vermont. But in the real world it is humanity that is blown up to planet-filling proportions. In the two centuries or so since the Industrial Revolution, human numbers have multiplied by eight and energy use and resource consumption have risen even faster. People now consume, control, or destroy almost 40 percent of the plant energy of the land and 25 percent of *all* plant energy, the ultimate source of food for all animals and almost all organisms.

As a result, quite unwittingly and ignorant of the consequences, we are reshaping earth, replacing forests with farmland, farmland with wasteland, filling rivers, lakes, and seas with sediments and pollutants, unbalancing the atmosphere, subtracting species and draining gene pools, changing climate, indeed changing earth faster, perhaps faster than it has changed ever before. We are revising creation. All this destructive effort has brought affluence to a mere fifth of the world. The remaining four-fifths struggle against increasing squalor and misery. We are alienating the whole of nature to meet human needs, yet human needs are not being met. Simply put, our relationship with earth is unsustainable.

So here we are, we citizens of Middlebury and other prosperous places, behaving like creatures from another planet. What should we do? We must make a daring change, a change to sustainability. But how?

The human species has gone through the agricultural and industrial revolutions. Now it is facing the need for a sustainability revolution. Unlike the transitions from hunting and gathering to agriculture, and from agriculture to industry, the transition to sustainability must be conscious. It requires people to behave as a global society, acting together, learning from each other's experiences, and sharing the planet's resources.

We need an explicit strategy. We need an explicit strategy because of the magnitude of the change required and because action is needed on many fronts and by many actors. The major problems, such as global climate change, population control, environmental degradation, and loss of biodiversity are strongly linked to each other. A solution to one requires solutions to them all. Moreover, no single group of actors, whether governments or citizens, can succeed by acting alone. A strategy is needed to ensure that all concerned take the priority actions on all the main fronts and that their actions are mutually supportive and move toward a common goal.

A strategy is also needed because we must learn from each other as we go along. The change to sustainability is fraught with uncertainty. We know some of the steps that must be taken but by no means all of them. We know roughly the direction in which to go. But we have only a vague idea of what the destination will be like. A strategy will help us learn from each other's experience and apply the lessons widely and rapidly.

Caring for the World is offered as such a strategy for sustainability. It is being prepared by three organizations, the World Conservation Union (IUCN), a union of more than five hundred governments and nongovernmental organizations in well over one hundred countries, and thousands of scientists and professionals concerned with management of natural resources; the United Nations Environment Programme (UNEP), which has been described as the environmental conscience of the UN, and the World Wide Fund for Nature, the largest private fundraising organization for conservation.

Caring for the World builds on the *World Conservation Strategy* (1980) and the report of the World Commission on Environment and Development, *Our Common Future* (1987). It has been prepared in collaboration with the Food and Agricultural Organization of the United Nations,

UNESCO, the World Bank, and most other agencies for the United Nations, as well as many other organizations and professionals around the world. The aim of *Caring for the World* is to set out the strategic directions and priority actions that will enable people to make the change from unsustainable to sustainable ways of life. Sustainability is a characteristic of a process or state that can be maintained virtually indefinitely. It implies living within earth's regenerative capacity:

> The living world is a like a cake
> That can replenish what you take.
> Leave certain bits and eat the rest slow—
> Back the cake is sure to grow.
> Sustainability means that you
> Can have your cake and eat it too.

Thus sustainable development means improving the quality of human life while using a constant level of physical resources.

Eight principles of sustainability are listed in *Caring for the World*, but I think that here they can be reduced to six:

1. Limit human impact on the living world to a level that is within carrying capacity.
2. Conserve the conditions of life—maintain the stock of biological wealth. This requires: maintaining life-support services (the ecological processes that sustain the productivity, adaptability, and capacity for renewal of lands, waters, air, and all life on earth); maintaining the variety of life in all its forms (different species, the genetic variability of each species, and the variety of different ecosystems they form); and ensuring that all uses of renewable resources are sustainable.
3. Minimize the depletion of nonrenewable resources.
4. Aim for an equitable distribution of the benefits and costs of resource use and environmental management.
5. Promote long-term economic development that increases resource productivity and natural wealth.
6. Promote values that help achieve sustainability.

In addition, six strategic directions are proposed for making the change to sustainability:

1. Transform attitudes and practices. The value system that dominates the world today is utilitarian. It is centered around physical growth and consumerism and is largely market driven. These attitudes may be changing. Demands are increasing to protect nature and to share responsibility for future generations. At the same time, there is a growing respect for values that are based on a deeper understanding of the human species as part of nature and that give greater weight to nonmaterial benefits. However, these changes are not happening fast enough or widely enough.

Sustainability calls for a fundamental transformation in how people behave. Changes in behavior can be assisted by laws and incentives, and by material help where needed, but the changes will last only if they arise from attitudes and practices based on a commitment to sustainability.

2. Build a global alliance. Thinking globally is not enough. We must act globally as well. Climate change, ozone depletion, and invasions of pollutions by air, rivers, and seas are making national borders increasingly irrelevant. Neither wealth nor sovereignty can protect us from them. True security can come only from safeguarding the earth. For this we need to build a global alliance based on the principle that all will have a role to play but those who have more must give more.

The earth is one. The atmosphere and the oceans are global commons, and rivers, seas, and migratory species are shared by many nations. Economic actions by upper-income countries affect the capacities of lower-income countries to protect ecosystems and to develop sustainably, which in turn affect the sustainability of the living earth on which everyone depends. Conserving the earth requires a true partnership among the nations of the world.

But the people of the earth are not one. The most obvious disparity is between high-income communities, whose members use large amounts of energy and raw materials, and low-income communities, whose per capita resource consumption is modest. Because of wasteful and excessive consumption, high-income communities often degrade the environments of other communities, as well as their own. Heavily populated, low-income communities often degrade their own environments as they struggle to meet their basic needs.

An effective strategy for sustainability depends on the partnership of all communities, which in turn depends on an equitable sharing of responsibilities. The rich and powerful must do more than the poor and vulnerable. High-income countries, like Canada and the United States, must lead the way by:

- Reducing their resource demand—their consumption of energy and materials.
- Maintaining, in an exemplary fashion, ecological processes and the variety of life in their own countries. Biological diversity is not confined to the tropical rain forests of Brazil. It is all around us.
- Greatly increasing capital and technical assistance for sustainable development by the lower income countries. This includes reducing their debt and improving their terms of trade. Due largely to the crippling burden of debt there is a net flow of funds from low-income countries to high-income countries. This must be reversed.

We need to: retire enough Third World debt to restore economic progress; make development assistance compatible with sustainability; develop true partnerships among nongovernmental organizations of all countries; and adopt a global Convention on Sustainability. To expand from nationalism to globalism, we need a treaty—an international understanding of what sustainability means to the peoples of the world, and of the rights and duties of nations to secure and to share the earth.

3. *Empower communities*. Individual and community action is the ultimate basis for national and global sustainability. The local level—the level of the individual, community, and locality—is where ecosystems are conserved or destroyed, needs are met or frustrated, and ecological, social, and economic factors are integrated. Governments and international organizations can help people to develop sustainably but cannot accomplish sustainability for them. Communities and individuals need to be empowered to adopt sustainable lives through education and training, and by increasing their control of the resources they use, their participation in conservation and development projects, and their influence on decisions that affect them.

We need to promote Primary Environmental Care. This concept means self-help by individuals and communities in an effort to enhance their environments and to achieve sustainable development. Effective Primary

Environmental Care requires secure access to resources and information and full participation in decisions. We must enable communities to prepare and adopt their own local strategies for sustainability.

4. *Integrate environment and development*. The economy has always been an integral part of the environment. Much of the unsustainability of current development is due to treating them as if they were separate. To bring us back down to earth we need to integrate environment and development in policy, law, economics, institutions, research and information, and monitoring.

Actions to achieve this include: make sustainability a primary goal of economic and development policies, reflecting that goal in budget and investment decisions; establish the commitment to sustainability in law; make liable those who deplete biological wealth or damage the health of people or ecosystems; include environmental costs in the prices of energy, raw materials, and manufactured goods; use economic instruments to provide incentives for sustainable action; make free access to environmental information a fundamental right; incorporate changes in environmental health and the stocks and flows of natural wealth in national accounting systems.

5. *Stabilize resource demand and population*. Human impact on the biosphere is the product of the number of people multiplied by how much in energy and raw materials each person uses or wastes. Energy use per person is a key factor. Someone whose only energy sources are food and fuelwood will use fewer resources and have a lesser environmental impact than someone who can afford to buy large amounts of fossil fuels and electricity.

The forty-two countries with high levels of energy consumption per person contain a quarter of the world's population but account for more than three-quarters of its consumption of commercial energy. The 128 countries with relatively low levels of energy consumption per person contain three-quarters of the world's population but account for less than a quarter of commercial energy consumption. On average, a person in a "high-energy" country consumes eighteen times more commercial energy than a person in a "low-energy" country. As a result, people in "high-energy" countries cause much more environmental destruction than people in "low-energy" countries.

Limiting human impact on the biosphere, therefore, calls for:

· Reducing per capita resource consumption levels in countries where they are relatively high.

· Assisting countries where they are low to develop and adopt technologies that use energy and resources efficiently.

· Stabilizing world population as soon as possible.

The main actions we propose are: replace taxes on income and sales with taxes on consumption of energy and other resources; help lower-income countries adopt resource-efficient systems; join and support "green consumer" movements; triple the supply of family-planning services in the 1990s; improve the status of women and maternal and child health care; direct development to increasing the security and incomes of the poorest families.

An important difference between our proposal to tax energy and other resources and other proposals along these lines is that we propose that this tax be instead of existing income and sales taxes. A resource consumption tax does not have a chance if it is in addition to these taxes. So we suggest that energy/resource taxes be phased in and progressively replace income and sales taxes.

6. *Conserve the variety of life*. Of the three requirements for maintaining the stock of biological wealth—maintaining life-support services, conserving the variety of life, and using renewable resources sustainably—conserving the variety of life is singled out because it is the most fundamental and the most neglected strategic direction.

It is easy to understand that life-support services like maintaining the equilibrium of the atmosphere and maintaining renewable resources are important and that more effort is needed to maintain them. But biodiversity includes millions of species that may have no measurable economic value. Accordingly—and experience with the World Conservation Strategy bears this out—a special effort must be made to focus on biodiversity, or it will quickly become an afterthought.

To conserve the variety of life we need to: join and uphold international agreements to conserve bio-diversity; provide communities with strong incentives to conserve bio-diversity; support management of biological resources by local communities and encourage the formation of grassroots organizations for that purpose; build conservation of bio-diversity into development policies and practices; complete and maintain a comprehensive global system of protected areas; and estimate the contribution of biodiversity to the national economy.

The Need for a World Ethic of Sustainability

Returning now to the first strategic direction—transform attitudes and practices—we propose three sets of actions: promote a world ethic of sustainability; increase awareness of the need for sustainability and incorporate environmental education in all formal and informal education programs; and improve and impart the skills and knowledge needed for sustainable development. Of these actions, promotion of a world ethic of sustainability is central.* A world ethic is needed because individual actions now combine to have global effects. For the first time in human history, we must be individually conscious of our impacts on the planet. Since value conflicts and competition for scarce resources are worldwide in scope, the ethical principles to resolve them must be shared globally.

The absence of an adequate ethic of sustainability is a major factor responsible for our failure to meet basic human needs, for growing inequities and the loss of freedom in the use and enjoyment of nature, for the loss of diversity and integrity of cultures and ecosystems, and for the destruction of the capacity of the biosphere to support future generations. The need for a world ethic of sustainability—an ethic that helps people cooperate with one another and nature for the survival and well-being of all individuals and the biosphere—could not be greater.

We propose two actions to promote a world ethic of sustainability. The first is to form an international coalition for a world ethic of sustainability. This would:

- Foster dialogue on the relationship between human rights and environmental duties.
- Explore ways to bring the world ethic of sustainability to the forefront of decision making by governments, intergovernmental bodies,

*Here I would like to thank Professor Ron Engel, who chairs the World Conservation Union's Ethics Working Group. This area of the strategy, its quality, and the prominent place it occupies in the strategy are entirely due to his leadership.

educational institutions, businesses, other organizations, communities, and individuals.

- Work with governments and jurists to incorporate the world ethic of sustainability in a Universal Declaration and Convention on Sustainable Development.

The second action is to establish a system to alert the world's people to serious breaches of the world ethic of sustainability and the *World Charter for Nature*. The activities undertaken by such a system would be analogous to those carried on by Amnesty International in the field of human rights. It could provide a powerful forcing mechanism by mobilizing world public opinion to demand higher standards of behavior toward the earth by governments, corporations, and others. The system would be operated by an independent international nongovernmental organization set up expressly for the task.

Consistent with the analogy with Amnesty International, the proposers of the *World Charter for Nature* originally considered that the organization might be called "Amnesty for Earth." However, "amnesty" in all definitions means pardon for past offenses (real or imagined). Since the organization would be concerned with preventing and correcting "offenses" against the earth—not by the earth—Amnesty for Earth would seem to be an unsuitable name. An alternative is Advocates for Earth. The advantages of the latter name are that it describes accurately what its members would be doing, and each member could be called an "advocate for earth." Its disadvantage is that "advocate" has a connotation of lawyer (and in some languages that is its primary meaning).

Amnesty International has an outstanding record of bringing abuses of human rights to world attention and of improving the conditions and obtaining the release of prisoners of conscience and political prisoners. This success may be attributed to five key characteristics:

- It is a genuine grassroots organization.
- It has a carefully focused mandate, concentrating on a limited number of objectives.
- It is independent and impartial.
- Its members adhere to well thought-out procedures.
- Its actions are based on thorough research.

I believe that Amnesty for Earth or Advocates for Earth would have to share these characteristics to be as effective. For example, just as the goal of Amnesty International is to secure throughout the world observance of the *Universal Declaration of Human Rights*, so the goal of Amnesty for Earth should be to secure throughout the world observance of the *World Charter for Nature* and the proposed Universal Declaration and Convention on Sustainability. And just as Amnesty International avoids dissipating its efforts by restricting itself to three objectives—helping prisoners of conscience; ensuring fair and prompt trials for all political prisoners; and opposing the death penalty and torture—so Amnesty for Earth should restrict itself to a few clear objectives, especially enforcing a world ethic of sustainability.

CHAPTER 8

Faith and Community in an Ecological Age

Steven C. Rockefeller

Fig. 8. St. Francis, who has been named the Patron Saint of Ecology, offers a valuable example, from within the Western tradition, of reverence, friendliness, and tender concern for all life. David Barten's (1942–) wooden statue St. Francis of Assisi *(1983; polychromed by Doris Karsell, 1922–) draws both on the techniques of medieval sculpture and on the American folk idiom. Through such an approach, he suggests the pertinence of ancestors like St. Francis to the environmental challenges of the artist's own culture and time. (St. Francis in the Fields Episcopal Church, Harrods Creek, Kentucky.)*

HUMANITY'S search for its spiritual center and the quest for a new way of life in harmony with ecological stability are converging today. The environmental crisis cannot be addressed without coming to terms with the spiritual dimension of the problem, and the spiritual problems of humanity cannot be worked out apart from a transformation of humanity's relations with nature. The integration of the moral and religious life with a new ecological worldview, leading to major social transformations, is a fundamental need of our time. This essay seeks to clarify these issues, giving special attention to a historical perspective on environmental problems and possible solutions.

Moral and Religious Dimensions of the Problem

At the outset it is useful to discuss briefly the moral and religious dimensions of the environmental crisis that faces human culture. Population planning, alternative methods of economic production and distribution, state regulation, new forms of political participation, and technological innovation are of critical importance in reversing the degradation of the environment. More fundamentally, however, what is needed is a commitment, so wholehearted as to be justly termed religious in quality, to a new ecological worldview involving a dramatic transformation of the moral values and basic attitudes that govern life in the industrial-technological world. Only such a radical shift in values and attitudes will bring about and sustain the full range of required social changes.

Moral thinkers have traditionally been concerned with the values that govern community life. They explore reflectively the idea of the good life and the concepts of right and wrong as applied to human conduct. They are often interested in the growth of the self, but they are especially concerned with the rights and duties that human beings have in their relations with one another. Some moral codes, such as one finds in the ancient Hebrew Torah, set forth guidelines pertaining to the treatment of animals and the land. The social and moral traditions that have been dominant in the West, however, have not involved the idea that animals, trees, or the land in their own right, as distinct from their owners or their Creator, have moral standing. Only a few saints and reformers have taught that people have direct moral responsibilities to nonhuman creatures. Usually, animals and ecosystems have been viewed primarily as means rather than as ends in themselves. They are thought to possess instrumental value but not intrinsic value. Consequently, a person's relations with them falls outside the sphere of right and wrong, unless these relations also affect the welfare of other persons. It is this widely shared outlook that is now being challenged, raising the possibility of a major revolution in the moral thinking of Western civilization.

Few people would deny that the environmental crisis necessarily confronts society with a number of new critical choices that concern values. For example, people must choose between enjoying clean air and a lifestyle that is heavily dependent on fossil fuels. They must choose between pure water and cheap methods of waste disposal. They must choose between the economic benefits of cutting ancient forests and the survival of wilderness and creatures like the spotted owl. These choices involve questions about the quality of life and the values people prize most. However, in what sense are these questions pertaining to values moral matters that fall within the sphere of what is appropriately labeled right and wrong conduct?

There are two basic approaches to arguing that it is morally wrong to degrade the natural environment or to abuse nonhuman creatures.[1] One line of reasoning is anthropocentric and utilitarian, contending that protecting the environment and nonhuman creatures is a matter of enlightened human self-interest and intergenerational responsibility. For example, the traditional American conservation movement, of which Theodore Roosevelt's Forest Service director Gifford Pinchot was perhaps the outstanding early spokesperson, points out that it is self-defeating to consume natural resources in a fashion that is not sustainable. It is harmful to the economic well-being of the human community, and therefore, morally wrong. Furthermore, even if the present generation does not suffer the consequences of the unsustainable use of resources, it is morally irresponsible to deny future generations the benefit of these goods. A similar kind of argument is made about preserving places of great scenic grandeur that refresh and inspire the human spirit or natural areas valued for the recreational opportunities that they afford.

Regarding the rights of animals, with support from the influential philosopher of natural rights, John Locke (1632–1704), modern humanitarians concerned about the abuse of animals have argued that when a society allows cruelty to animals, it develops a tolerance for cruelty that encourages the abuse of humans.[2] Reforming humans in their treatment of animals is a necessary part of improving human moral character and the quality of relations between humans. Therefore, the argument goes, it is appropriate to teach that it is morally wrong for humans to mistreat nonhuman sentient beings even though these beings do not possess the intrinsic value of persons.

Even though these lines of thinking may limit application of the concept of moral obligation to the relations between human persons, they do lead to a sense of respect for nature and a certain restricted notion of animal rights. Particularly when they are reinforced with a scientific ecological appreciation of interdependence, they begin to expand the human sense of community, opening the door to a new form of moral awareness. When this new expanded form of moral awareness is fully developed, it leads to a biocentric rather than an anthropocentric philosophy of respect for nature.

In a biocentric approach, the rights of nature are defended first and foremost on the grounds of the intrinsic value of animals, plants, rivers, mountains, and ecosystems rather than simply on the basis of their utilitarian value or benefit to humans. It may be argued in the language of Western liberal democracy, for example, that just as humans of all races and both sexes possess inherent worth and have an inalienable right to life, liberty, and the pursuit of happiness, so other life forms have a right to life, freedom from human oppression, and a habitat that offers them opportunity for well-being.

The ability and willingness to recognize in other beings intrinsic value and moral standing grows with the sense of community. A number of philosophers and social scientists have pointed out that the history of human morals is a story of a developing sense of community that begins with the family and tribe and then gradually extends outward embracing the region, the nation, the race, all members of a world religion, and then all humanity.[3] The sense of community involves an awareness of kinship, identity, interdependence, participation in a shared destiny, relationship to a common good. It gives rise to the moral feelings of respect and sympathy, leading to a sense of moral obligation. The British philosopher Jeremy Bentham (1748–1832), for example, argued for a sense of moral community with animals, including protection by the law, on the grounds that humans can identify with animals as sentient beings who experience pain and pleasure. A moral person, argued Bentham, is by definition one who is committed to minimizing the suffering of all sentient beings and to fostering happiness among them wherever possible.[4] There are an increasing number of environmentalists who believe that unless humanity's sense of moral community is dramatically expanded to include not only future generations of humans but also other life forms and even ecosystems, there is little hope of achieving the alteration in human behavior necessary to deal with our environmental problems. The environmental crisis is a crisis in our understanding of and commitment to community.

Having made these observations about the moral dimensions of the environmental crisis, it is important also to consider in what sense this crisis has a religious dimension. First of all, if a major transformation is to occur in human values and behavior, it will have to involve a concern and faith that is religious in nature. A religious concern is one that is a matter of fundamental controlling interest to a person. A person is religiously concerned about those values which he or she regards as essential to fulfillment in the deepest sense. To be religiously concerned about a set of moral values is to have faith in those values, to trust them as true guides to enduring well-being and peace. A moral faith that is religious in nature has a unifying effect on the personality, focusing and releasing energy.

A genuinely religious faith is not a product of the ego or of a particular act of self-conscious rational choice. It springs from the deeper center of the self and involves the whole personality—feeling, thought, and will. In biblical language this deeper center is often called the heart. Emergence of a moral faith that is religious in quality involves awakening on the level of the heart and a change of heart. This awakening of faith is not so much the result of grasping liberating moral truth as of being grasped by the truth. In the act of faith the whole self is possessed by the object of faith, surrendered and committed to the truth. Such religious concern and faith are essential to major personal and social transformations. If the earth is to be saved, there must be a new faith in a vision of the good that includes environmental ethics, a faith that is religious in nature in this broad sense.

Religious experience involves an encounter with the sacred, an intuition of the awesome and wondrous mystery in the power of being. The experience of the sacred is of critical importance in the transformation of human attitudes toward nature and the awakening of a new moral faith. An appreciation of the miracle of life and of the beauty and mystery in the being of animals, plants, and the earth as a whole must become so intense as to generate a keen sense of the natural world's sacredness. Dostoevsky's Father Zossima speaks about such a religious appreciation of nature: "Love all God's creation, the whole earth and every grain of sand in it. Love every leaf, every ray of God's light. Love the animals, love the plants, love everything. If you love everything, you will perceive the divine mystery in things. Once you perceive it, you will begin to comprehend it better every day."[5] A fresh awareness of the sacred values in nature fosters respect and moral responsibility.

The life and work of Wangari Maathai, a former professor of veterinary medicine and director of the Green Belt Movement in Kenya, demonstrates the nature and power of an ecologically oriented faith experience. Wangari Maathai has succeeded in organizing tens of thousands of women and children in an effort to reverse the process of deforestation and to empower local communities. Under her leadership over seven million trees have been planted, and the women of Kenya, who are responsible for well over half of their nation's food production, have begun to work together in new liberating ways. In an interview, Maathai gave the following explanation as to why she had become so involved as an environmental leader.

> I just have something inside me that tells me there is a problem and I must do something about it. So I am doing something about it. I think that it is what I would call the God in me. All of us have a God in us, and that God is the spirit that unites all life—everything that is on this planet. And it must be this voice that is telling me to do something. And I am sure it is the same voice that is speaking to everybody . . . who seems to be concerned about the fate of the world.[6]

Maathai's sense of being possessed by a divine imperative associated with an inner voice arising from "the spirit that unites all life" is the kind of faith experience that leads to an enduring ecological awakening.

This discussion of an expanding sense of community and a religious faith in environmental ethics would be incomplete without recognizing that a solution to the problems created by humanity's exploitation and abuse of nature is intimately bound up with the effort to overcome problems of social oppression and injustice. It is sometimes argued that humanism and environmentalism, the struggle for social justice and the quest for eco-justice, are movements with contradictory goals. Some fear that the worldwide struggle for human rights and economic development for the poor is being slowed by ecologists and animal rights activists. However, a growing number of social thinkers now recognize that the struggle to liberate people and the movement to liberate the larger biosphere are interdependent.

This view is sound for two reasons. First, ongoing deterioration of the environment and nonsustainable consumption of resources destroys the possibilities for long-term economic advance in the undeveloped world and denies opportunity to all future generations. Second, the ideas, attitudes, and values that lead to the abuse of nature are very closely related to those that cause the oppression of women, children, the poor, minorities, racial groups, and religious groups. From this point of view, the cause of democratic humanism is one with the cause of environmentalism.

The critical issue is to understand, criticize, and transform the attitudes of greed, hatred, and domination in whatever context they appear, seeking to get at the deeper social and psychological roots of the egocentricity at work in these attitudes. The contemporary search by humanity for its spiritual center acquires clear direction today when it is focused on the effort to find creative alternatives to a way of life that emphasizes egoism, having, and subjugation. The spiritual quest of our time is for a faith and ways of living that at once liberate the self and the other, creating an authentic community with nature as well as among people.

The Shape of the Problem in Historical Perspective

By considering some of the ideas that have historically encouraged a lack of respect for nature and attitudes of exploitation, it is possible to gain further insight into the larger spiritual problems of humanity as well as the spiritual significance of the environmental crisis. In western culture there are at least three principal sources of beliefs that reinforce negative attitudes toward nature. They include the Judeo-Christian tradition, Greek philosophy, and the intellectual traditions associated with the emergence of modern science and the Cartesian and Newtonian worldview. It should be kept in mind throughout the following brief exploration of these ideas and attitudes that there are also important currents of thought associated with Judaism, Christianity, Greek philosophy, and modern science that emphasize different values and that are supportive of respect for nature, but they have not exercised a controlling influence during the modern period. In other words, within all these Western traditions there are strong correctives to the contemporary degradation of nature. As the twentieth century draws to a close, these corrective forces are beginning to assert themselves, giving rise to a global reform movement.

Certain aspects of the biblical tradition can be seen as generating anthropocentric, dualistic, hierarchical, and patriarchal ideas and attitudes that are problematic from an ecological as well as a democratic perspective.

For example, texts in Genesis and elsewhere indicate that God and nature are clearly separated, leaving nature without inherent sacred significance. In many biblical texts God is characterized as an all-powerful patriarchal figure who pursues conquest and destruction as well as loving kindness and peace. Alone among all living creatures, human beings are believed to have been created in the image of God, and they are consequently given the charge to "subdue" the earth and "have dominion" over it (Genesis 1:28). Moreover, the purpose of the creation of the universe is the establishment of a kingdom of God on earth by and for human beings. Nature is created to be the stage upon which this drama is played out, and it is primarily a means to this divine-human end. Since nonhuman life forms and the rest of nature exist in a realm separated from God and are not created in the image of God, they possess instrumental but not intrinsic value. Consequently, they are not part of the moral community. A hierarchy of God, humanity, and nature constitutes the structure of reality. In addition, men are viewed as having rightful dominion over women on the grounds that the reproductive functions of women render them more closely related to the earth and hence weaker in matters of the spirit, as symbolized by the seduction of Eve by the serpent.

These ideas, which found currency in influential Jewish and Christian traditions, were merged over time in the popular mind with another set of problematical ideas developed in ancient and classical Greece. Plato created one of the great visions of the spiritual quest, but he taught that there is a radical separation between the spiritual world and the realm of nature, and between the soul and the body. Medieval Christianity under the strong influence of this tradition developed a hostile attitude toward the body, which in the Augustinian tradition came to be viewed as an obstruction both to purity of heart and to clarity of mind. Augustine did not believe that the body and the natural world were inherently evil. He did believe, however, that given the corruption of the human will as a result of original sin the body and the natural world constitute a source of perilous temptation, leading men and women to attachment to the lesser good rather than to the absolute good, God. He taught that the task of the soul is to liberate itself from the influence of the body and to escape from the natural realm into a divine spiritual world. In addition, many Christian thinkers in this tradition came to believe that nature itself had been reduced to a fallen state as a consequence of humanity's fall. As a matter of practical attitude, Augustinian Christianity fostered in the minds of the faithful dualistic ways of thinking about God and nature, soul and body, that gave a special urgency to the biblical injunctions to subdue and control nature. One needed to master nature in order to find salvation as well as to survive. There is, of course, an element of truth in this teaching, but it is one-sided and overemphasizes the negative. Also, in this outlook Western patriarchal society continued to rationalize the domination of women by men, arguing that women were by nature more deeply attached to the body and the earth than men.

Other Greek intellectual traditions also reinforced human anthropocentric attitudes in relation to nature. Ptolemaic astronomy regarded the earth as the center of the universe, encouraging a human-centered vision of the cosmos. Aristotle had a deep appreciation of biology and developed a profound naturalistic philosophy, but he also imagined nature to be governed by a fixed hierarchy of essences and to have its supreme controlling end in the development of the power of reason, the essence of human nature. In and through the work of Thomas Aquinas, Aristotle's hierarchical way of understanding the world was merged with Christian hierarchical thinking. Aristotle's philosophy involved a deep appreciation of unique human capacities and his teleological interpretation of nature recognized in all things a certain inherent value, but hierarchical modes of thought can easily be used to foster the idea of nature as a mere means in a universe that exists primarily to serve the needs of humanity or a certain class of humanity.

This kind of perspective on nature was reinforced by another Greek tradition quite separate from Aristotelian philosophy. One of the most significant Greek and Roman cosmologies was the atomistic materialism of Democritus, Epicurus, and Lucretius. According to this point of view, reality is constituted by discrete, indestructible, material atoms, and the entire universe is constructed out of their purely mechanical relations. This atomistic, materialistic, and mechanical view of the world won wide support in European civilization with the development of seventeenth-century science and philosophy, which interpreted nature in exclusively mechanistic and mathematical terms.

With novel and forceful arguments, René Descartes (1596–1650) reconfirmed the classical Greek dualism of mind and body, spirit and nature, and unwittingly gave fresh encouragement to Western egocentricity. In and through his famous *cogito ergo sum* ("I think, therefore I am"), he asserted the self-evident reality of the ego as an atomic, self-conscious, rational mind and will and gave it the central governing position in the world as he described it. His sharp distinction between the ego and its world involved the notion that the world of nature is properly understood as a great machine, made up of matter in motion and governed by the laws of mathematics. Animals and plants were to be understood basically as little machines. This view of nature found its most complete scientific expression in the work of the great English physicist, Sir Isaac Newton (1642–1727).

Descartes along with Francis Bacon (1561–1626) was one of the early prophets of the power and promise of science. His idea of the ego brought into full consciousness the rational powers and methods of thought that generated the modern scientific revolution. The emergence of the Cartesian ego in Western culture involved development of an extraordinary capacity to objectify the world, to turn everything into an object or thing to be analyzed, manipulated, and controlled. The powers of the Cartesian ego and the mathematical-mechanistic view of nature generated the industrial revolution. The development of the Cartesian ego also contributed to the Western drive toward the independence of the self, encouraging new demands for intellectual, social, and economic freedom. When aspirations for democratic social change were joined with the forces of industrialization and technology, the idea of progress was born, leading to a new secular faith that emerged as a powerful social force in the late eighteenth and nineteenth centuries. Mastery over nature and democratic social transformation promised not only material well-being but a new way to salvation, the realization of an ideal society of freedom, justice, and equal opportunity on earth.

As history has revealed, however, the science and philosophy of the seventeenth and eighteenth centuries, enlightened as it was in many respects, did not provide an entirely sound basis for realization of progress and the perfection of human freedom. First of all, the Cartesian and Newtonian worldview widened the gulf between the human spirit and nature and gave powerful support to the tendency in Western culture to view nature as an object only, a mere means, which has led to attitudes of reckless conquest and exploitation. Francis Bacon gave expression to this outlook when he triumphantly proclaimed that the new science would soon make "Nature with all her children" the "slave" of humankind.[7] As a result of these developments, in the course of the eighteenth century the dualistic, hierarchical, anthropocentric attitudes at work in Christianity acquired new secular forms of expression in the emerging modern industrial society.

Though the idea of the Cartesian ego and Newton's mechanistic physics gave humanity new powers of control, they also precipitated a spiritual crisis. Men and women found themselves feeling strangely alienated from nature, which science portrayed as a great mindless machine operating without any controlling purpose. The naturalistic explanations of the new science discredited the old supernaturalistic interpretation of events. The rejection of the idea of divine intervention in the natural order coupled with the growing secularization of society, made God seem increasingly remote and unreal. Many thoughtful people developed an eerie sense of existing in a vast indifferent and meaningless universe that cared nothing for truth, beauty, and goodness. The Cartesian dualism of self and world coupled with an atomistic psychology seemed to create an unbridgeable gulf between subject and object, intensifying feelings of isolation. The sense of estrangement deepened as industrialization and urbanization created unhealthy and stressful mechanized social environments. In this new world, people also found the forces of mechanization having a dehumanizing effect on human relations. They began to think of themselves and others merely as things, objects, cogs in a great industrial wheel. Beginning with the Romantic movement in the late eighteenth and the early decades of the nineteenth century, Western culture has reacted in diverse ways against the dualism, rationalism, atomism, materialism, and mechanism at work in the Cartesian and Newtonian worldview, but the sense of spiritual disorientation and alienation has persisted into the late twentieth century.

The most influential eighteenth- and nineteenth-century visions of the ideal society to be realized by science, democracy, and progress unleashed powerful revolutionary forces of liberation, but they were limited in scope in certain important respects. The early visions of this ideal new world were inspired by the idea of freedom and are of enduring moral significance, but they were still encumbered by attitudes of domination and exploitation. They were thoroughly anthropocentric visions that did not include nonhuman life forms in the moral community, and it was not just animals, plants, and forests that were denied moral standing, but humans as well. There was no place in the moral community for enslaved persons until the mid-nineteenth century, and the doors to full citizenship for women were not opened until the twentieth century. Furthermore, through much of the nineteenth and twentieth centuries the industrial and political leaders of Europe and America were inclined to view the rest of the world, including its people, as primarily an opportunity for imperialistic adventures and exploitation. There were radical democratic and ecological movements in Europe and America that challenged the dominant outlook. In the eighteenth century, for example, the Shakers taught the abolition of slavery, the equality of the sexes, and "justice and kindness to all living beings."[8] However, only through a slow and painful process would such visions be integrated into mainstream thinking.

There is a passage in C. G. Jung's *Memories, Dreams, Reflections* that points to the deeper problems in Western culture that were intensified with the development of Cartesian rationalism and scientific materialism. Jung tells the story of a conversation he had with a Pueblo Indian, Ochwiay Biano (Mountain Lake), during a trip to New Mexico in the 1920s.

> "See," Ochwiay Biano said, "how cruel the whites look. Their lips are thin, their noses sharp, their faces furrowed and distorted by folds. Their eyes have a staring expression; they are always seeking something. What are they seeking? The whites always want something; they are always uneasy and restless. We do not know what they want. We do not understand them. We think that they are mad."
>
> I asked him why he thought the whites were all mad.
>
> "They say that they think with their heads," he replied.

> "Why of course. What do you think with?" I asked him in surprise.
>
> "We think here," he said, indicating his heart.
>
> I fell into a long meditation. For the first time in my life, so it seemed to me, someone had drawn for me a picture of the real white man. It was as though until now I had seen nothing but sentimental, prettified color prints. This Indian had struck our vulnerable spot, unveiled a truth to which we are blind. . . . What we from our point of view call colonization, missions to the heathen, spread of civilization, etc., has another face—the face of a bird of prey seeking with cruel intentness for distant quarry. . . . All the eagles and other predatory creatures that adorn our coats of arms seem to me apt psychological representatives of our true nature.[9]

Jung sensed that Western culture in its dominant modes of expression had lost touch with the heart, the deeper center of the whole self, and with the human impulses to sympathy, care, and intimacy. The idea of the self—especially in the case of popular images of the successful male—had become narrowly identified with reason and the will-to-power within the framework of the Cartesian-Newtonian worldview.

The development of the Cartesian ego in Western culture has involved a one-sided development of the self. The Cartesian ego has achieved wonders. It is responsible for humanity's awesome scientific accomplishments, and it has enabled the human race to enter an industrial/technological phase of development with its many advantages as well as dangers. The difficulty has been that the growth of the rational powers and independence of the ego, the separated "I," has not been counterbalanced by a corresponding growth of compassion, the sense of community, and the capacity for intimacy, which spring from the heart. The split between the head and the heart described by Jung is a modern expression of a very old problem symbolized by the biblical story of the Fall (Genesis 2–3).

The myth of the Fall describes the emergence of self-consciousness and the ego in the human life cycle. The growth of the ego makes possible the separation of subject and object, knowledge, and autonomous choice, and in this sense it is a step toward freedom and self-realization. It also involves the disruption of an original harmony of the self with nature, other persons, God, and itself. All but the most severely abused children have some direct experience of this primal unity during their early years when, as Wordsworth has expressed it, the world was "apparelled in celestial light." The heart's deepest longing is for a recreation of the lost harmony. This cannot be achieved by regression, a return to an infantile world. The self achieves this end by integrating the head with the heart and by employing the ego's powers of intelligence and will in building liberating relationships and a community of freedom, justice, and peace. Wholeness, enduring meaning, and union with the divine spirit are found in and through relationship and community. For a variety of complex reasons, however, including ignorance, pride, and fear reinforced by social conditions, human beings are led to perpetuate their "fallen" state. To a significant degree their capacities for creative liberating relationships lie buried within them and go undeveloped. Instead of putting the emphasis in society on the human resources for communicating, cooperating, sharing, loving, and creating, the dominant tendency has been to seek fulfillment in an ego centered orientation that stresses having, competing, controlling, and using.[10] This is, of course, a generalization. American democratic life at its best, for example, does not fit this description, but the promise of democracy as well as the meaning of ecology has yet to be realized in full.

The split between the head and the heart in Western culture has found expression in the divisions between science and faith, fact and value, spirit and nature, ultimate meaning and everyday life, the sacred and the secular, the individual and the community, the self and God, male and female, and oppressor and oppressed. These unhappy separations leave modern culture haunted by a sense of estrangement, by anxiety about meaninglessness, and by the fear of death, nothingness. They underlie much of the widespread neuroses in contemporary society. Struggling with this situation in their "fallen," or ego-centered, state of mind, many people seek escape in consumerism, exciting diversions, drugs and alcohol, conformity, or fundamentalism while others are driven to pursue a quest for power. The sad fact is that learned, clever, and powerful though it be, the Cartesian ego, which has created the wealth and power of modern society, can neither find in itself nor create in the larger world lasting peace. The one-sided development of the self resulting from a preoccupation with the powers of the ego has created a particularly dangerous situation. Having eaten further of the Tree of Knowledge, leading to industrialization and the discovery of atomic energy, but not having effectively integrated intelligence with compassion and developed the needed wisdom in education and social planning, modern humanity has become a threat to its own survival as well as to all life on the planet.

Prophetic Voices from the Past

The growth of democratic culture, which accompanied the development of science and technology, has led Western society to wrestle constantly during the last two centuries with the issue of expanding human rights and social justice. It was with publication of Rachel Carson's *Silent Spring* in 1962, however, that American democratic society at large began slowly to wake up to the environmental crisis and to the question of the rights of nature and eco-justice. In an effort to respond to the crisis, philosophers, theologians, and literary naturalists began to discover an array of rich traditions in Western culture upon which to draw in preparing the ground for ecological as well as a democratic social reconstruction. In addition, people in the West have grown increasingly aware of a variety of ancient Eastern traditions that offer different ways of envisioning the relation between humanity and nature and that suggest fresh approaches to environmental issues. A few examples drawn from the East and the West will illustrate the point.

Ancient Hinduism involved a high degree of world denial, but it also taught that human beings and nature are interdependent parts of a unified world, which is in its entirety a manifestation of Brahman, the undifferentiated Eternal One. In harmony with this outlook, Hinduism teaches a "reverence for nature and all forms of life," regarding the Earth as the Universal Mother who has nurtured the human race over millions of years of evolution.[11] The Jain tradition of India had a pessimistic view of life on earth and believed in a dualism of soul and body, but it also widened the moral circle to include all living beings including plants, contending that all living beings have souls and are of infinite value. Along with rigorous ascetic practices, the Jains pursue the path of *ahimsa*, or noninjury, in their relations to each and every living being.

The classical Buddhist doctrines of the interdependent arising of the universe (*dharma-dhātu pratītya-samutpāda*), impermanence (*anitya*), and emptiness (*śūnyata*) constitute a vision of the universe as a self-creating, self-maintaining, evolving organic whole.[12] The universe is constantly generating an endless number of unique individual beings, but nothing exists in and of itself as a completely independent entity. All beings are "empty" in the sense that they do not have a permanent self that can exist by itself. All beings are interdependent in the sense that their existence and identity is dependent upon their relations to all the other things that make up the universe. Dualistic and atomistic ways of thinking about experience and reality are rejected in favor of an emphasis on the relationships between all things.

The practical implications of this Buddhist form of an ecological worldview are gratitude and respect for all things, each of which in its own unique way sustains the universe and the existence of all other beings. Wasteful consumption and exploitation are inconsistent with Buddhist ethics. The emphasis tends to shift from an exclusively anthropocentric to a more biocentric view. Furthermore, insight into the truth of the doctrine of emptiness, or no-self, dissolves the sharp boundaries between self and other, leading to identification with the life of the whole and the awakening of compassion for the suffering of other beings. The ethics of emptiness mean adopting an attitude of helping other sentient beings on the path to liberation insofar as one is able and of leaving them alone if one cannot be of help. Buddhism recognizes the uniqueness and intrinsic value of each sentient being and envisions an ideal world in which all can attain liberation and fully realize their individuality.

In China, this Indian Buddhist worldview was elaborated in the Hua-yen school of philosophy.[13] Chinese Buddhists also found themselves debating whether nonsentient beings like plants, trees, and rocks possess the Buddha-nature and are capable of enlightenment and liberation. Some argued that this is the case, and the debate spread to Japan. For example, Kukai, the early ninth-century founder of the Shingon sect of Buddhism in Japan, defended the idea that plants and trees have the Buddha-nature. He contended that the whole natural world is one with the Dharmakaya, that is, the living essence of the Buddha's teachings, which is emptiness. The influential twelfth-century poet Saigyō went so far as to maintain that human beings find the way to Buddhist liberation and the realization of emptiness in and through relation to the natural world. For many Japanese, nature itself is possessed of inherent sacred meaning and value and is a source of saving power.[14]

Quite apart from the influence of Buddhism, most Chinese philosophers, who have had roots in the Taoist tradition, tended to share a worldview that rejects anthropocentrism and the dualism of matter and spirit and of humanity and nature. The cosmos is viewed as a self-generating organic whole which is an open process of becoming. Each and every thing is an interdependent interacting part of the continuous cosmic process, which is called the "great transformation" (*ta-hua*).[15] The fundamental reality in this process is a vital force or power (*ch'i*), which embraces both the physical and spiritual aspects of the world. Rocks and trees as well as humans are manifestations of this vital force. It is omnipresent and preserves the interconnectedness of all things in the ongoing process of growth and transformation which is the universe. Humans are not superior to this natural process but integral parts of its functioning. The eleventh-century Confucian philosopher Chang Tsai writes: "That which fills the universe I regard as my body and that which directs the universe I regard as my nature. All people are my brothers and sisters, and all things are my companions."[16] Human persons find fulfillment in this organic view by dissolving their egoism and by living each in his or her own unique way in harmony with the great transformation, adjusting to and caring for the earth "as my body."

Some have questioned the value of Asian philosophies and religions in connection with the environmental crisis on the grounds that India, China, Japan, and other industrialized Asian nations are now causing the same kind of environmental damage as the West. However, it has been correctly pointed out that extensive environmental degradation in the East has to a large extent followed upon "the *intellectual* colonization of the East by the West," that is, the spread of the Cartesian and Newtonian worldview.[17] Given the universal tendency of human beings to get stuck in egoistic delusions, it should not be surprising that the Cartesian-Newtonian worldview and Western consumerism have had a seductive influence in the East as well as in the West. Eastern peoples struggle with problems of greed, hatred, and domination today just as all others. This proves nothing one way or the other about the truth or value of Buddhism or Taoism, or for that matter of Judaism, Christianity, or Islam. One thing is clear: spiritual awakening and transformation are difficult to achieve, and there are no easy answers or ready-made quick fixes available. The East as well as the West needs all the help that it can get in the form of spiritual insight and transformative methods. Traditions like Buddhism and Taoism remain vitally important sources of renewal available to Asian cultures. In addition, these Eastern traditions offer spiritual practices, insights into nature, and "a rich vocabulary of imagery, symbol, and metaphor" that can assist the West in developing its own new ecological worldview.[18]

There are also numerous voices and visions within the Western tradition that counsel respect for nature and challenge the kind of dualistic thinking that turns nature into a thing to be exploited as a mere means for satisfying human greed. The Hebrew Torah, for example, is to a large extent the story of a people and their relationship to the land. The land is understood to belong to God and to be valued by God as good. It is given to the people as a divine gift on the condition that they care for it as responsible stewards and live according to the divine law. There are many passages in the Torah that reveal ecological sensitivities and that counsel compassion for animals. The prophet Isaiah envisioned the animals as participating in God's final redemption.

One also finds in the Hebrew Scriptures a deep current of prophetic ethical thought that sees the meaning of history in liberation of the oppressed and the achievement of social justice. This biblical tradition has laid an ethical foundation upon which today the champions of animal rights and justice for nature are building. The existence of this prophetic tradition coupled with the Socratic spirit of rigorous rational criticism is one basic reason why a powerful environmental movement has been able to develop in Western societies. This should not be forgotten when biblical and Greek traditions are criticized on environmental grounds. In addition, the Hebrew prophets recognized that fundamental to realizing the divine purpose in history is a transformation of personality, a basic change in attitude, involving the awakening of an ethical faith that arises from the heart.

Jesus stands in this prophetic tradition. His own teachings are especially relevant to the current environmental as well as social situation because they involve a direct attack on a character orientation in life that is ego centered, emphasizing having, domination, and exploitation. The story of Jesus involves a great paradox: the Messiah of God, the most powerful of human beings, comes as one who is powerless from a worldly perspective. As Jesus explains it, the person who would be great must become a servant, and the one who would find his or her life must lose it. At the core of his ethical teaching is a prophetic call to seek God's kingdom before all else, loving the spirit of goodness, which is the very being of God, more than wealth, power, fame, and pleasure.

The path into the kingdom of heaven is the way of faith and purity of heart. Jesus explains what this means. Trust in God for your ultimate fulfillment. Thirst after righteousness. Clarify the eye of your mind. Abandon pride. Be humble. Empty yourself of greed. Free yourself of the lust that turns persons into things. Love your neighbor as yourself. Care for the poor and oppressed. Be trustworthy. Be a peacemaker. Do not get angry. Judge not. Be forgiving. Be merciful. Love your enemies. The emphasis in Jesus' teachings is on being good, as opposed to having goods.[19] To forget the self out of a heartfelt concern for the well-being and liberation of others is the practical meaning of the love of God and being good. God, the Eternal One, is the spirit of caring and sharing that creates community.

In the Christian tradition, St. Francis of Assisi, who is widely regarded among Christians as having come as close as any mortal to realizing the perfection of Christ, came to view the sun, the moon, the earth, and all living creatures as his brothers and sisters—as members of the moral community of God. Each creature praises God and shows forth the glory of God in his or her own unique fashion.[20] St. Francis urged his followers to respect water and fire and to treat worms and wolves with kindness. He also attacked the root of the problem when he counseled that "it is in giving that we receive." St. Francis is perhaps the most obvious historical Christian figure to whom Christians might turn for the purpose of an ecological reconstruction of Christian ethics, but there are many other Western theologians and philosophers whose thinking is significant as well.

John Macquarrie's Gifford Lectures (1983–1984), which have been published under the title *In Search of Deity*, are a good example.[21] Macquarrie is representative of a number of Christian theologians who, prompted by problems with classical theism including its tendency to create a radical separation of God and nature, have discovered a rich Western panentheistic tradition with roots in Neo-Platonism and Christian mystics like Dionysius the Areopagite (c. 500 C. E.), Hildegard of Bingen (1098–1179), and Meister Eckhart (1260?–1329). *In Search of Deity* traces the development of this tradition from Plotinus (205–270), to Alfred North Whitehead (1861–1947), and Martin Heidegger (1889–1976). Panentheism, or dialectical theism as Macquarrie prefers to call it, seeks to counterbalance the one-sided emphasis on the transcendence of God in much traditional theism with a deep appreciation of the immanence of God. It pursues a middle way between pantheism, which too simplistically identifies God and the world, and classical theism, contending that all things are in God and God is in all things. The effect of panentheism is to locate matter, the body, and the earth within—instead of outside—the Divine, and to locate God at the center—instead of on the periphery—of reality and the everyday world. In this way it tends to break down the dualism of the spiritual and the natural, soul and body, the religious life and the common life, heaven and earth.

One finds in the thought of a great Islamic theologian and mystic such as al-Ghazālī (1058–1111) a vision of God that is similar in many respects to certain forms of Western panentheism. There are in the Islamic tradition other spiritual resources that can be employed to counter the degradation of the environment. The Quran, for example, warns against the extremes of anthropocentrism:

Assuredly the creation
Of the heavens
And the earth
Is greater
Than creation of humankind;
Yet most people understand not.

Quran (XL:57)

Sufism, the Islamic mystical practice of inner illumination, and parts of the *Sharī'ah*, the holy law of Islam governing the outward life, support a sacramental view of the natural world and respect for nature. Most fundamentally, Islamic spiritual practice is concerned with the achievement of purity of heart and an attitude of peace and compassion. Commenting on the Islamic concept of Jihād, holy war, one contemporary Sufi master, Bawa Muhaiyaddeen, writes: "For man to raise his sword against man, for man to kill man, is not holy war. True holy war is to praise God and to cut away the enemies of truth within our own hearts. We must cast out all that is evil within us, all that opposes God. This is the war that we must fight." Bawa Muhaiyaddeen further comments: "We must learn to wash away our separations and become one again. That is true Islam."[22]

At the core of the great religions has been a quest for the divine that has sought in a variety of ways to expand and elevate the self beyond a narrow egoism. Moses set his people free, but he sought to humble them with faith and to socialize them with the law. The Buddha taught a doctrine of no-self. Over the centuries the attack of religious prophets against egoism has led to an evolutionary process involving an ever-expanding moral vision and sense of community. There are, of course, within all the religions strong social and political forces that have repeatedly tried to arrest the process and to identify the religious vision with various fixed forms of tribalism, sectarianism, and nationalism. However, deep within the creative religious consciousness there is a tendency to break free of divisive structures and attitudes in a quest for the fullness of divine truth and goodness. The traditions just considered manifest that tendency. The environmental crisis provides humanity with the challenge of working out the full implications of this quest.

Twentieth-Century Visions of the Greater Earth Community

As the twentieth century progressed, a number of philosophers and scientists developed fresh visions of the world and of the good life that anticipated and prepared the ground for the shift in attitudes and values that has become a matter of broad interest at the end of the century. In this regard the contributions of Albert Schweitzer, Martin Buber, and Aldo Leopold are of particular significance.

Albert Schweitzer (1875–1965) stands out among Christian philosophers for his radical expansion of the idea of the moral community. During the First World War he found himself trying to understand what lay at the root of the moral failures that had plunged Western civilization into a horrifyingly self-destructive conflict. If industrial-technological society was to survive and flourish in the future, Schweitzer concluded, it would have to undergo a major ethical transformation, involving an awakening to the sacredness of all life. The most immediate fact of human consciousness, he argued, is the will to live. A healthy spirituality begins with conscious affirmation of one's own will to live, recognizing existence "as something unfathomably mysterious." Affirmation of life means reverence for life and devotion to full self-realization. Schweitzer goes on to point out that a thinking person is also conscious that his or her life exists "in the midst of life which wills to live," and whoever reflects on this situation "finds a compulsion to give to every will to live the same reverence for life that he gives to his own. He experiences that other life in his own."[23]

These reflections led Schweitzer to a new definition of a moral person.

> He accepts as being good: to preserve life, to promote life, to raise to its highest value life which is capable of development; and as being evil: to destroy life, to injure life, to repress life which is capable of development. This is the absolute fundamental principle of the moral.
>
> A man is ethical only when life, as such, is sacred to him, that of plants and animals as that of his fellow man, and when he devotes himself helpfully to all life that is in need of help.[24]

Schweitzer knew well enough that a human being cannot completely avoid harming and destroying other life forms; he counseled, however, that an ethical person guided by "Reverence for Life" tries to minimize suffering wherever possible. Unless the modern world awakens to this greater sense of ethical community, Schweitzer contended that there was no hope of improving the relations between people and achieving lasting peace.

Much of the spirit of Schweitzer's reverence for life is also found in the thought of the Jewish philosopher Martin Buber (1878–1965), whose book *I and Thou* (1923) was destined to have a major impact on Western religious thought in general. In certain respects, Buber was more radical than Schweitzer, for he argued that "every thing and being" should be respected as a Thou, an end in itself, and not merely as an it, a means to be used for ends external to its own being.[25] "In all the seriousness of truth, listen," wrote Buber, "without It a human being cannot live. But whoever lives only with that is not human."[26] Buber described I-thou relations with trees, rocks, soil, and tools as well as human persons. Furthermore, he argued that one encounters God, the Eternal Thou, not as a being apart from the world, but in and through I-thou relations with the things and beings that make up the world. I-thou encounters give life inherent and ultimate meaning, and they alone are the key to lasting community and peace, the only sure path away from the horrors of twentieth-century history. Buber's thought involved a world-affirming philosophy that sought to overcome completely the dualism of the sacred and the secular, divine meaning and everyday life: "At each place, in each hour, in each act, in each speech the holy can blossom forth."[27] The fundamental task of humanity is to hallow this earth in and through the spirit of I and thou. Buber did not fully develop a theory of environmental ethics, but the implications of his philosophy for a transformation of human ethical attitudes toward nature as well as toward all human beings are abundantly clear.

The thought of Schweitzer and Buber calls into question the Western industrial-technological idea of the way to progress. They do not reject the idea of progress but point out that if there is to be progress, it must rest on a new ethical and religious foundation and be concerned first and foremost with the quality of a people's relations with each other and the larger earth community. They recognize that progress in the area of social justice and human liberation is intimately related to a revisioning and transformation of humanity's relations with nonhuman life forms and the larger natural world. In this connection they emphasize that the critical issue is a fundamental shift in attitudes and values. The urgent question is whether a person's character orientation in relation to the world is governed first and foremost by an I-it or I-thou approach or some variation on this theme.

In figures such as Henry David Thoreau and John Muir, the American tradition has its own early prophets of a respect for nature that includes appreciation of ecological values and the intrinsic worth of nonhuman life forms. However, the forester and nature philosopher Aldo Leopold (1886–1948) is widely regarded as the father of contemporary biocentric environmental ethics, because he integrated ethical thinking with the new ecological science. Before turning directly to Leopold's thought, it is helpful to consider briefly the approach of scientific ecology, and also that of twentieth-century physics, which has influenced the development of ecology.

In both contemporary physics and ecology, an organic and holistic view of nature has replaced the mechanical and dualistic outlook of the Cartesian and Newtonian worldview. Reflecting on new developments in these two sciences, the philosopher J. Baird Callicott writes: "energy seems to be a more fundamental reality than natural objects or discrete entities—elementary particles and organisms, respectively. An individual organism, like an elementary particle is, as it were, a momentary configuration, a local perturbation, in an energy flux or 'field.' "[28] Regarding the holistic concept of nature, the new physics and ecology make it clear that individual organisms can only be fully understood when appreciated as parts of the energy field or matrix in which they are formations in constant process of change. "Contrary to the object-ontology of classical physics and biology in which it is possible to conceive of an entity in isolation from its milieu—hanging alone in the void or catalogued in a specimen museum—the conception of one thing . . . necessarily involves the conception of others and so on, until the entire system is, in principle, implicated."[29] In other words, a thing is constituted by and cannot be conceived apart from its relationships. Nature is a unified whole in the sense that all things are interrelated, systematically integrated, and mutually defining. The principle of interdependence is as fundamental as the principle of individuality in nature.

A holistic and relational view of the human self suggests that in a real sense the whole earth is a person's extended body and the consciousness of the individual is a focal point of the earth's emerging self-consciousness. Such an awareness creates the possibility of transforming egocentricity and anthropocentrism into environmentalism. The ecological worldview emerging in Western culture is convergent with important aspects of the worldviews associated with Buddhism, classical Chinese philosophy, and some forms of Christian mysticism and panentheism. In addition, the scientific shift from classical atomism and Newtonian mechanistic materialism to the new physics and ecology complements the social movement from feudalism to democracy. The ecological worldview encourages a certain egalitarian appreciation of the intrinsic value and contribution of all things, and it deepens the democratic understanding of community, clarifying the way in which all individuals are reciprocally interconnected and dependent on development of the common good. Liberal democracy has often been afflicted with an excessive individualism, which reflects the prevailing ideas at the time of its emergence. Ecological ways of thinking check the perpetuation of a faulty atomistic psychology and egocentric delusions.

Aldo Leopold was one of the first thinkers to develop the ethical implications of ecology. His ethical outlook is set forth most clearly in a well-known essay entitled "The Land Ethic," which appeared in his *A Sand County Almanac* (1949). He closely relates ethics to the human sense of community:

> All ethics so far evolved rest upon a single premise: that the individual is a member of a community of interdependent parts. His instincts prompt him to compete for his place in the community, but his ethics prompt him also to co-operate (perhaps in order that there may be a place to compete for).
>
> The land ethic simply enlarges the boundaries of the community to include soils, waters, plants, and animals, or collectively: the land.[30]

Leopold further argues that "land . . . is not merely soil; it is a fountain of energy flowing through a circuit of soils, plants, and animals. Food chains are the living channels which conduct energy upward; death and decay return it to the soil."[31] So defined land constitutes what Leopold termed "the biotic community."

In his ethical vision the well-being of the biotic community as a whole is the supreme good, and the human being is understood first and foremost as a citizen of this community rather than as a conqueror of it. Land is viewed as "the collective organism" rather than "the slave and servant" of humanity. As a citizen of the biotic community, a human being has obligations to the land and its nonhuman members that go beyond those dictated by economic self-interest. As interdependent members of the community, other life forms have a "biotic right" to exist, since they are "essential to its healthy functioning." In this sense, other life forms are ecological equals in their relations with humans. What is needed, argued Leopold, is emergence of a new "ecological conscience" that views appropriate land use as an ethical rather than merely an economic problem. He recommended the following criterion as a guide: "A thing is right when it tends to preserve the integrity, stability, and beauty of the biotic community. It is wrong when it tends otherwise."[32] With the holistic emphasis of his land ethic, Leopold was less concerned than Albert Schweitzer and animal rights activists about the rights of individual creatures as distinct from the species. This difference of emphasis has led to an ongoing debate on how to balance holism and a respect for individual rights.

Faith and an Ethic of Sustainability

Prior to the 1960s very few Americans were interested in Schweitzer's "Reverence for Life," Leopold's "Land Ethic," or the implications for relations with nature of Buddhism, Taoism, panentheism, Buber's I-thou philosophy, and the new physics and ecology. However, during the decade of the sixties, which saw the dramatic rise of movements directed against racism, sexism, and militarism, the environmental movement, too, began to take form. Contemporary civilization has entered a period when a critical shift in worldview and values is in the making. At the time of the first Earth Day in 1970, the environmental movement was by and large a white middle-class movement of Americans and Europeans inspired by scientific research and their Western moral and aesthetic sensibilities. Twenty years later it has become a worldwide movement actively supported by peoples from all races, classes, and religions. Environmentalism has moved into the political mainstream, and its advance is increasingly seen as interconnected with that of other social liberation movements.[33] The forces of resistance remain very powerful, but in all the major sectors of society, groups demanding change are gaining in influence.

In the course of the past two decades, philosophers and theologians have taken up the challenge of a new environmental ethics in earnest. The *World Charter for Nature*, which was approved by the General Assembly in 1982, was the first widely supported international effort to codify the emerging new values (see Appendix). With the emergence of the concept of sustainable development as an organizing principle that integrates environmentalism and economics, efforts are now under way to formulate "a world ethic of sustainability." Proposals for such an ethic are being developed by the International Union for Conservation of Nature and Natural Resources and the United Nations Environment Programme, and other groups.[34] Intensive efforts are being made to shape the ethics of sustainability into an Earth Charter to be finalized at the Earth Summit being organized by the United Nations Conference on Environment and Development for Rio de Janeiro in June, 1992.

One might call what is happening a new kind of "Great Awakening." It is in part a response to the acute environmental degradation, social fragmentation, international conflict, and psychological stress that mark the times. On the positive side, it is the result of the convergence of a variety of social, scientific, international, ethical, and religious movements that all point to the urgent need for a transformation in our way of life based upon a new sense of interdependence, community, and responsibility that is global and intergenerational, including women and men, all races and religions, all life forms, and the Earth as a whole. People are led to embrace this new way of imagining the world and living by traveling diverse pathways, which are often interconnected. They include diverse religious visions, moral democracy, various holistic philosophies, the new physics, the science of ecology, reverence for life, deep ecology, the practice of I and thou, feminism, and the ethics of sustainable development.

In the final analysis the issue before the human race is an ancient problem that has appeared in a new social context in an especially dangerous way. The new social context is the industrial-technological world with its nuclear capabilities, which has been created by the Cartesian-Newtonian worldview and the scientific method. The old problem is the one symbolized by the story of the Fall of Adam and Eve. It involves the failure of human beings to integrate the separated "I," the rational ego, with the affections and sympathies of the heart, the deeper center of the self, and to create social conditions that promote the balanced growth of all. Education is of critical importance in addressing environmental problems, but a concept of education that does not encompass this basic problem is not fully adequate.

It is the intimations and promptings of the heart that remind human beings of a better way, of ideal possibilities forgotten and not realized. It is through imaginative vision and the power of faith that people let go their egoism, overcome their fears, and commit themselves to the ideal, giving themselves over to the methods of personal and social transformation that can actualize the ideal. Faith has the power to take the self beyond itself to a new self. The emergence of a liberating faith involves a change of heart in the sense of an awakening to the needs and capacities of human nature for growth and for cooperating, sharing, loving, and creating. The venture of faith is a matter of wholehearted decision for which all are ultimately responsible.

The awakening of faith, however, cannot be mechanically engineered and controlled. Nevertheless, faith is a natural part of the growth process and the flowering of human personality. Society can tend the conditions of its growth by creating a liberating social environment and by celebrating the values of authentic comprehensive community in art and ritual. Individuals can nurture faith within themselves by listening to the promptings of their hearts and, in Socrates' language, by caring for their souls. Today caring for our souls importantly includes what Wangari Maathai, the Kenyan environmental leader, described as listening to the voice within us of "the spirit that unites all life" and then responding with creative action.

In this new ecological age of developing global community and interfaith dialogue, the world religions face what is perhaps the greatest challenge that they have ever encountered. Each is inspired by a unique vision of the divine and has a distinct cultural identity. At the same time, each perceives the divine as the source of unity and peace. The challenge is to preserve their religious and cultural uniqueness without letting it operate as a cause of narrow and divisive sectarianism that contradicts the vision of divine unity and peace. It is a question whether the healing light of religious vision will overcome the social and ideological issues that underlie much of the conflict between religions. It is a twofold challenge, reflecting the present planetary situation. Just as all societies struggle today to build a world community and to transform their way of living in relation to nature, so the religions are faced with the need to create community among themselves and to build a new sense of community with nature. Insofar as the world's religions are faithful to their purist visions of the divine harmony and take to heart the ecological idea of global interdependence, they will be able to help the human race at this critical moment in the evolution of its consciousness.

The awakening of a faith in ecological, as well as democratic, values and ideals is essential to the long-term success of the environmental movement. Also, the ethics of ecology projects a vision of global community and an idea of the liberated self that is fundamental for any creative religious faith in the twenty-first century. Without the awakening to such a faith, the freedom that human beings have managed to achieve may well eventually result in the destruction of the possibilities for life on earth. The urgency associated with the environmental crisis, however, means that it has a special power to make us see beyond the limitations of our ignorance, pride, and fear and to awaken in us a new ecological faith, which will have diverse forms but a common ethical core. The roots of such a faith lie deep in ancient traditions, but it is the destiny of this and future generations to bring this faith into full flower in the light of the new knowledge of the interdependence of the whole earth community.

NOTES

1. In the following discussion of environmental ethics, I am particularly indebted to Roderick Frazier Nash, *The Rights of Nature: A History of Environmental Ethics* (Madison, Wis.: University of Wisconsin Press, 1989), 1–86, 121–60.
2. Ibid., 14–19.

3. Ibid., 1–7, 11–12, 42–48, 52–54.
4. Jeremy Bentham, *An Introduction to the Principles of Morals and Legislation*, ed. Laurance J. LaFleur (New York: Hafner Publishing Co., 1948), 310–12. See also Nash, *Rights of Nature*, 23–24.
5. Fyodor Dostoevsky, *The Brothers Karamazov,* trans. Constance Garnett and ed. Ralph E. Matlaw (New York: Norton, 1976), 298.
6. Wangari Maathai as quoted in "Race to Save the Planet," Program 10: "Now or Never," Annenberg CPB Collection, produced by NOVA, WGBH, Boston. See also Jane Perlez, "Nairobi Journal: Skyscraper's Enemy Draws a Daily Dose of Scorn," *New York Times International*, 6 December 1989.
7. As quoted in William Leiss, *The Domination of Nature* (New York: Braziller, 1972), 55.
8. See outline of founder Ann Lee's (1736–84) teachings on a nineteenth-century Shaker broadsheet, The William Benton Museum of Art, University of Connecticut, Storrs, Conn.
9. C. G. Jung, *Memories, Dreams, Reflections* (New York: Pantheon Books, 1963), 247–48.
10. For a very helpful exploration of the contrasting character orientations under discussion, see Erich Fromm, *To Have or To Be?* (New York: Harper and Row, 1976).
11. Karan Singh, "The Hindu Declaration on Nature," in *The Assisi Declarations: Declarations on Religion and Nature*, World Wildlife Fund 25th Anniversary, 29 September 1986 (London: Waterloo Printing Co., Ltd., 1986), 17–19.
12. See Francis H. Cook, *Hua-yen Buddhism: The Jewel Net of Indra* (University Park: Pennsylvania State University Press, 1977), 14–15, 30.
13. See, for example, Cook, *Hua-yen Buddhism*.
14. William LaFleur, "Saigyō and the Buddhist Value of Nature," in J. Baird Callicott and Roger T. Ames, eds., *Nature in Asian Traditions of Thought* (Albany: SUNY Press, 1989), 183–208.
15. Tu Wei-Ming, "The Continuity of Being: Chinese Visions of Nature," in Callicott and Ames, eds., *Nature in Asian Traditions*, 67–78.
16. Chang Tsai, "Western Inscription," in Wing-tsit Chan, trans. and comp., *A Source Book in Chinese Philosophy* (Princeton, N.J.: Princeton University Press, 1969), 496. See also Tu Wei-Ming, "The Continuity of Being," 69–74.
17. J. Baird Callicott and Roger T. Ames, "Epilogue," in Callicott and Ames, eds., *Nature in Asian Traditions*, 279–84. See also Eugene C. Hargrove, "Foreword," in ibid., xiv–xvii.
18. J. Baird Callicott and Roger T. Ames, "Introduction," in Callicott and Ames, eds., *Nature in Asian Traditions*, 17.
19. Donald S. Lopez, Jr., and Steven C. Rockefeller, "Introduction," in Donald

S. Lopez, Jr., and Steven C. Rockefeller, eds., *The Christ and the Bodhisattva* (Albany: SUNY Press, 1987), 9–11.

20. St. Francis of Assisi, "The Canticle of Brother Sun," in Eloi Leclerc, *The Canticle of Creatures: Symbols of Union*, trans. Matthew J. O'Connell (Chicago: Franciscan Herald Press, 1977), xvii–xviii.
21. John Macquarrie, *In Search of Deity: An Essay in Dialectical Theism* (London: SCM Press, Ltd., 1984).
22. M. R. Bawa Muhaiyaddeen, *Islam and World Peace: Explanations of a Sufi* (Philadelphia: Fellowship Press, 1987), 44, 98.
23. Albert Schweitzer, *Out of My Life and Thought: An Autobiography*, trans. C. T. Campion (New York: Holt, 1933), 145–58.
24. Ibid., 158–60.
25. Martin Buber, "The Way of Man According to the Teaching of Hasidism," in *Hasidism and Modern Man*, trans. Maurice Friedman (New York: Harper and Row, 1958), 127, 173–74.
26. Martin Buber, *I and Thou*, trans. Walter Kaufmann (New York: Scribner's, 1970), 85.
27. Buber, *Hasidism and Modern Man*, 31.
28. J. Baird Callicott, "The Metaphysical Implications of Ecology," in Callicott and Ames, eds., *Nature in Asian Traditions*, 51. See also Stephen Sterling, "Towards an Ecological World View," in J. Ronald Engel and Joan Gibb Engel, eds., *Ethics of Environment and Development: Global Challenge and International Response* (London: Belhaven Press, 1990; rpt. Tucson: University of Arizona Press, 1990), 77–86.
29. Callicott, "Metaphysical Implications of Ecology," 59–60.
30. Aldo Leopold, "The Land Ethic," in *A Sand County Almanac* (Oxford: Oxford University Press, 1949), 239.
31. Ibid.
32. Ibid., 237–64.
33. Lester R. Brown, "The Illusion of Progress," in Lester Brown, *et al.*, eds., *State of the World 1990: A Worldwatch Institute Report on Progress Toward a Sustainable Society* (New York: Norton, 1990), 4, 12–16.
34. The current efforts to formulate a "world ethic of sustainability" are briefly discussed in the introduction to this volume, and a summary of the kind of ethical guidelines being considered in this regard is presented by the editors of this volume in the introduction to Robert Prescott-Allen's "Caring for the World," chapter 7.

CHAPTER 9

"Keeping Faith with Life": A Dialogue

Steven C. Rockefeller, moderator

Fig. 9. The Chinese landscape painter Shih Lu (1919–1982) was a dissident who at one point in his career took refuge in the wilderness. His depiction of pines on one of China's sacred mountains, Pines on Mt. Hua *(1970s), expresses his admiration for the perseverance of the spirit of freedom and resistance in a time of oppression. The artist inscribed the following poem on his painting. (The Metropolitan Museum of Art, New York. Gift of Robert Hatfield Ellsworth, in memory of La Ferne Hatfield Ellsworth, 1986.267.351.)*

I love the pines of Mount Hua,
Tall, noble, solemn, and dignified.
Their thrusting trunks vie with the sun and moon.
Resisting cold winds through the years,
They shake their arms at the sky-scraping ridge
And hold high their heads, like striding blue dragons.
They support the clouds forever,
Without taking flight to the heavens.

On the third day of the Middlebury symposium, the contributors to the symposium and this volume participated in a panel discussion. This chapter contains an edited version of that dialogue.

STEVEN ROCKEFELLER: Where do we go from here? We have heard many inspiring addresses, and some powerful visions have been sketched out for us. At this juncture it would be helpful if we inquired further into how in our own personal lives and in society, we might go about the process of reconstruction that is the challenge. Therefore, I would like to ask this question: "What do you, as individuals, do in your life in order to internalize the environmental values that we have been talking about? And what advice can you give to this audience about what they might do in their personal lives?"

SEYYED HOSSEIN NASR: I would answer the question in two parts. First of all, I believe "ecological values" are not values which are rooted in the ecological situation; they are values which descend from God into the world of nature. Therefore, the interiorization of "ecological values" is really the primary goal of my life, namely to interiorize myself, to try to reach the inner realm within my being wherein resides what Christ called "the kingdom of God," the inner reality where the divine compassion, according to the Quran, descends. And from there to try to live in such a way that this interiority is reflected in the world about me—to be able to see the face of God in His creation.

Second, I try in a more externalized way to reflect in what I write and what I say the significance of seeing nature as the domain of the sacred. I try to live with nature in such a way that these inner values are reflected in nature and then in a very modest way to be a model for those who would follow the same way.

AUDREY SHENANDOAH: I would begin by saying that I probably have the advantage because I was raised this way from as early as I can remember. I didn't learn the things that I know, the things that I talk about, the things that I am trying to pass on to the younger generation, from any school of education, and I am not a scholar. I was very fortunate in being brought up by my grandmother and my grandfather, who were elders when I came to them as an infant. So I heard all of this from the beginning of my time. I heard about the respect for the Earth, the respect for all of creation, the respect for other people.

There have been many times within my people's history and during my time, when we have had confrontation with our white brothers, many times with the government or neighboring communities. My grandmother never taught me, so I never learned in our house, to hate or to dislike or to find fault. She always told me to give thanks for those that I know and for what I know and for what I see.

Every morning we were not allowed to wash our faces with warm water. Every morning we washed our faces with cold water. I still do to this day. But we should give thanks for this water the minute we open our eyes, because water is life. We should also give thanks for one more day. That is what we give thanks for in our ceremonies; one more day to enjoy this creation and the gifts that are so bountiful, if we would only recognize them, acknowledge them, and use them rightly.

So I was learning all my life these things that I'm trying to pass on to the young people. I learned about the respectability of women within our society. I learned about the very prestigious position our women hold in our society. But I was also learning what we must do with our own lives in order to receive that respectability. And so I try very hard consciously to live the way that I was taught to live—with this respect for everybody and every other living thing, for our mother, the Earth, our grandmother, the Moon, and especially the water.

I work a lot with children, so what I do has to be something that is going to remain a good image in the eyes of these young people. If I am going to teach them anything, if I expect them to learn anything from me, then I have to show them an example by doing and being what it is I'm trying to teach and pass on to them.

My advice can only be that it has to begin with every individual. It's going to come from your inner self. We are responsible for our own space, however big our own space is. We have to keep peace within that area that we are responsible for. No matter where you move—no matter what the condition.

J. RONALD ENGEL: There are many ways that I would think about answering this question. But I suppose finally it comes down to a recurrent theme, and that is locating ourselves comprehensively. And once one begins to locate oneself directly in response to the fate of this earth, one finds oneself making new covenants. At the deepest level, I have been making new covenants.

That means I have become increasingly aware internally of the thousands and thousands of human beings who have lived on this earth and who have loved this earth and who never dared think that it might come to an end or to the destructive consequences that now are so evident to us. I have tried to make a covenant with those people, with that past, with their hopes. This finally involves one's relationship to life itself. And so a term that has become very important to me personally is keeping faith with life.

Most concretely, in terms of specific actions this means making promises. I have made promises to myself, promises to my wife, promises to my children, promises to my community—that my work and my life will take this as its center to the degree possible. It's that sense of inner centering based in a covenant and faithfulness to the adventure of life on this earth.

SALLIE MCFAGUE: I wish it were the case that I could say with Elder Shenandoah that I have been learning about my proper place in nature all my life. I haven't been. In my own case, it came about as a result of a conversion several years ago, and I use that word advisedly. I mean by a conversion a basic transformation of sensibility in which I began to see my place differently, our place differently, than what I had perceived it to be before. Rosemary Ruether, in one of her books, talks about the need for us to convert our minds and hearts to the earth. I think that is exactly what needs to be done. In other words, a basic *metanoia*, repentance, a turning around. A conversion is a total kind of thing after which one sees differently. The scales have fallen from our eyes, and we perceive our place and the place of others in a different way than we did before. All of us need to experience a basic conversion, or at least most of us who haven't been brought up in traditions that have encouraged that from the beginning. How to place oneself in such a position so that that will occur is part of what needs to be done.

Once that did occur for me, it seemed crucial not just to be able to spend my free time on this issue, if it is *the* issue of the twenty-first century that we've been saying it is. If it's not just another issue but a matter of life and death for our species and other species, then it isn't sufficient simply to work on it in one's spare time. One needs to realign one's work and central focus in directions that are going to be on the side of helping solve this problem rather than contributing to it. My advice is that you look at your own life, your own interests, your own abilities, your own concerns, and ask yourself how they can be made central to the planetary agenda. Every one of us must contribute in a central way to that.

I have been encouraged lately reading polls that show that the young people in the country see the ecological deterioration as the central concern facing us. Maybe it's because that's the world that they all are going to live in. My generation will probably be able to make it through without an enormous amount of discomfort. That's a very crass way of putting it, and to me it is not sufficient because the thought of one's own death is surely nowhere nearly as horrible as the thought of the gradual death and decay of our planet. In fact, I think most people can contemplate their own death with more ease, if they think that other generations will come and that the joy and happiness of a diverse, healthy planet will be there after they have gone.

ISMAR SCHORSCH: I would like to be very concrete. For our personal lives it is necessary to be just a little less self-indulgent. I don't use a computer. Wendell Berry has a beautiful essay on why he doesn't use a computer. I like the physicality of writing with a pen and a pad. I think we've lost something in the machinery of even the typewriter and certainly the computer—it's all become so much more mechanistic.

The immediacy of expressing oneself has been shattered by the intervention of machinery. I still listen to music on LP's. I have managed to ignore the revolution since the LP, and I'm not sure that I'm the poorer for it. I've subjected my family to an automobile without an air conditioner, and I drive my secretary to distraction because I turn the air conditioner on in my office so little. I don't wish to be virtuous; I merely suggest that a little bit less self-indulgence isn't going to hurt us and probably will make a solid contribution to the environment.

And then I think it's most healthy to think in terms of the future in reference to our kids. Most of us aren't built to think abstractly or to identify with global values. But we sure do identify with our children, and that's what I think this is all about. What kind of world will our children live in? And if we relate these global, abstract, overwhelming problems to our children, I think our motivation will be profoundly intensified.

In the next generation, the challenge will be to learn to live with far greater inner resources. Success must not only be defined by income or by how many homes one has or by how many cars one drives. Those are the norms of the society from which we come, and they are not very helpful toward the future. Success has to be defined in far more elusive spiritual terms: the concerts that we hear, the gardens that we plant, the poems that we write, the books that we read, the walks that we take. It is that inner dimension of humanity that is ultimately going to make it possible for us to live in harmony with our environment.

I believe that that is the goal of a liberal arts education. That's why our young men and women are at universities and colleges across this great land, not to prepare for law school or medical school but to acquire the skills so that once they become lawyers or doctors, they will be able to enrich their inner lives. They will not have to work twenty hours a day in order to earn hundreds of thousands of dollars; they will find meaning in earning less but having more time for their children, more time for themselves, more time for the community. That is the challenge of this generation. We've not done very well facing that challenge. We did not have the compulsion of reality to push us in that direction. But I am convinced that the cultivation of the inner life is one of the direct contributions that individuals in the next generation can make to addressing the deterioration of the planet.

THE DALAI LAMA: I believe that when each individual realizes there are long-term negative consequences, each will see that it is important to work to remove these negative consequences, according to one's own profession or line of work. In my own case, I am a Buddhist monk, and whenever I have time to explain about Buddhism, I try to explain the importance of taking care at least of one's own environment. For example, when I visit a Tibetan settlement, I always tell them to plant trees, trees with fruit, you see, that kind of thing. And then also as human beings, we need a certain discipline, contentment, awareness, and what I call a sense of universal responsibility. These things I myself am trying to practice. Some people are technicians, some economists, some teachers, and according to one's own profession, there is this possibility to contribute to the common aim. So one utilizes in the maximum way one's own potential.

ROCKEFELLER: In the course of the symposium, much has been said about cooperation among all of us as human beings and between the world religions. Is it realistic to think that in the immediate future the world religions will be willing and able to cooperate in addressing environmental problems? If we look at what's happening in Ireland, Lebanon, the West Bank, the Persian Gulf, and many other parts of the world, religion is caught up in very destructive social conflicts.

NASR: Let me say, first of all, that the situation of religion in the West is not the same as the situation of religion in other places. We have been presented with a very eloquent discourse by Professor McFague concerning the deconstruction of certain images of the Christian tradition. For me, as one who happens to be a Muslim, who knows a little about Christianity, and has lived a long time with Christians—even for me, it is very difficult to understand how if Christ said, "Father, forgive them for they know not what they do," that now the image of Father will somehow have to be changed. With the Christ, was that wrong, or was the Apostle wrong in noting it down in the Gospels? This represents a very difficult point for a non-Christian from the other side of the world to understand. I only give this as an example. This means that within Christianity there is, if not a crisis, then certainly a tension of views which is not at all the same as in other religions.

Nevertheless, this having been said, I believe that within every religion those who are attracted to the question of nature and ecology are those very ones who are also attracted to what is most inward and, therefore, universal within religions. I remember a metaphor given by the late Cardinal Pignierdolli who was the Vatican Cardinal in charge of non-Christian religions. He and I used to discourse together about Christianity and Islam. He always used to say that geographically it's the boundaries of countries which meet, but in the world of religion it's only the hearts of different religious countries which meet. Now, it's precisely because only the hearts of religions really meet, that those within various religious traditions who are interested in the ecological problems are those also who are interested in the inner dimension of these religions.

And therein lies a great deal of hope. Usually those religious authorities within various religions who are interested most in the legal, political, economic, and social facets of their religions and therefore less in nature, which has always been related in all religions to the inward dimension, to the mystical, to the contemplative dimension—those leaders will not be those who will be carrying out a profound dialogue religiously. I feel, therefore, that despite all the troubles there is definitely the possibility of both discourse and accord among religions as far as the environment is concerned.

ENGEL: I take a great deal of hope from what has happened in the last twenty-five years. Let us back up for a moment. Twenty-five years ago, the present conversation in which we are now engaged was not in existence. When Aldo Leopold wrote in the late forties that religion had yet to hear of the environmental concerns of this planet, he was correct, by and large. When Lynn White wrote his classic essay on the historic roots of the ecological crisis in the mid-sixties, that was considered a persuasive article, not because he was necessarily correct in his historical analysis, but because he had put his finger on an incredible apathy and lack of attention throughout all the historic traditions with regard to the environmental crisis.

In twenty-five years we have seen each of the major religions of the world, as well as the religions of many indigenous native people, reach a self-consciousness about the environment as an issue that must be addressed morally and theologically, that I don't think many of us in this room could have anticipated. That has taken several forms. It has meant in the first case a inward critique by scholars and spokespersons for these religions of the areas in which institutional religion has either failed to speak or has in fact been inadequate. The Christian faith certainly has been deeply involved in that, but it has by no means been the only one. After a period of what we might call critical evaluation there has come a period of strong affirmation and discovery of areas within each religion—doctrines, understandings, fundamental symbolic openings—which had been hidden and lost in the modern period. That has given many religious communities the confidence to speak.

Within the last ten years the religious communities have begun to speak across the secular/sacred divide. For example, at the Assisi Conference in 1986 when four or five of the great world religions met, for the first time they met with the world conservation community. So the dialogue between science and the secular disciplines and the religious community has begun in earnest.

About six or seven years ago I sat down and made a list of the number of projects that religious communities throughout the world had initiated to respond to the environmental crisis. And I came up with about forty or fifty that I knew about. Only five or six years later that list would be almost impossible to draw up, it would include so many initiatives throughout the world.

It seems to me that we are now on the verge of the next stage when we begin to get down to the serious business of identifying very concretely what the different religions have to learn from one another, what help they can offer to one another in correcting liabilities and deficiencies. It seems to me we are also on the verge of something the new version of the World Conservation Strategy calls for: an international coalition of communities of dedicated persons who are trying to speak to the environmental crisis out of religious convictions.

A few years ago at our IUCN General Assembly in Costa Rica, we had a workshop on environmental ethics and religion, and it was attended by groups throughout Latin America principally, although also from other parts of the world. It ended up being the workshop with the largest attendance. And what was so evident was the eagerness of grass-roots groups throughout the region to have a part in articulating the global environmental ethic for the twenty-first century. They were eager to be part of the shaping of this ethic. They wanted to learn from their elders, their traditions, the best philosophy, and the best science, but they also wanted this to be a people's movement. So I have great hope.

ROCKEFELLER: It's often thought that the environmental movement is largely a white, middle-class, Western movement, but I hear you saying that that is not the case any longer?

ENGEL: That is correct. In fact, if we look at the present world situation, it is more frequently the advanced industrialized nations that are holding back progress in this area. I will give the best example I can: the *World Charter for Nature*. It was adopted by the United Nations by a vote of 111 to 1. The one nation that voted against it was the United States of America. The nations that led the drafting and adoption, which began within the IUCN General Assembly in 1980, were African nations. I have seen in my own experience a drastic shift of actual initiative from the North. It's not only a matter that it's a larger community than the Northern, white, Western societies, but in fact the real initiative is being passed and has already probably been passed to the other nations of the world.

ROCKEFELLER: It is important to note in our deliberations, that the ecological conscience has really begun to awaken throughout the world and that the possibilities for international cooperation are very real.

SCHORSCH: I want to emphasize the significance of the emerging alliance of science and religion. What is most important today is that religionists are becoming sensitive to the environmental crisis and reaching out a hand to the scientists. We were not there first. We usually aren't on most of the important social issues. So what value is there in bringing the traditions to bear? We are the most effective public educators in the business. We do touch people's lives at existential moments when fear and anxiety overcome them. We do have a lot of power to transmit constructive values. If scientists point out the danger to the planet and we accept the challenge, I think that it is possible for us to reach a lot more lives than scientists could. Scientists are austere, remote personalities. It is the ritual of religion that touches most people in their daily lives. And if religion is mobilized in the effort to reduce our appetite and share more of our wealth, that I think is the most effective form of public education in which we can engage. That's what I think is beginning to happen across the world, an alliance of scientists and religionists prepared to martial all the resources for public education.

I feel that the key to this problem is not government regulation but human education. That's why the churches and the synagogues and the mosques are being mobilized for this effort. If we can change the attitude of individuals, then collectively over time we will change the way our society articulates its values and spends its money. That cannot be effectively regulated from above.

ROCKEFELLER: In the course of our symposium we distributed to students and faculty the IUCN/UNEP report *Caring for the World: A Strategy for Sustainability*, which is an attempt to set a global agenda for dealing with the environmental crisis including establishment of a new world environmental ethic. Is it necessary and possible to develop a universal ethic of sustainability that will be accepted by all the different religious traditions and by the many different cultures in the world, or is that an unrealistic or inappropriate goal?

MCFAGUE: This is a very important issue. There is a danger of a new universalism. We are at a time when women and a number of other people whose voices have not been heard—the so-called underside of history—are insisting upon different experiences, different voices, social forms different from the oppression under which they have lived. The danger is that we could move into a new type of the universalism which in the past has been a mask, usually, for white male European/American views, and so forth. So when asked about a universal ethic, I in turn ask: is this some kind of a watering-down of differences? Is it a refusal to entertain now the very different voices that are coming to us and that have not been allowed to be part of the discourse in the past? If it were that, then I think we would want to negate it. It's terribly important that the voices that have not been heard now be heard. This includes the voice of the creation. We have to speak for the voiceless ones who have not been able to speak.

On the other hand, as I looked over the "elements of a world ethic of sustainability" from the *Caring for the World* report, my concerns were allayed. Just to recall a few of the points: People are part of nature. Every life form warrants respect and preservation. All persons should take responsibility for their impacts on nature. People should treat all creatures decently. The needs of individuals in society should be met. Each generation should leave to the future a world that is at least as diverse and productive as the one it inherited and so forth. This to me is not a mask for a new kind of universalism. Nor do I think it waters down the differences among the various religions. I don't know any religion that was present at this symposium that would not support these points. If this is what an ethic of sustainability would be about, then it seems to me this would be a place where we could join in common agreement.

NASR: I would agree with Dr. McFague definitely. But I want to add one point. I do not think it's absolutely necessary to have a global agreement about these points in order to be effective. You always talk about the environmental problem being global. It might be global, but action is local. Human beings have to pay attention to what they're doing locally—to their house, to their local environment, their office, to their farm, whatever they're doing. They cannot act globally. I think we should be very careful not to become discouraged if we do not have a global consensus at the present moment. But as far as, certainly, the Islamic world is concerned, I would agree with Dr. McFague—all of these points would be accepted by all major representatives of the different parts of the Islamic world.

SCHORSCH: I think in many cases it's extremely difficult for different religions to cooperate, and a bit of pressure from the outside is very salutary. I think that's exactly what the ecological crisis offers. It is a common danger that may help religious groups and traditions transcend their particularity and begin to raise their sights to the welfare of mankind in general. So whereas religions may not have the internal resources to overcome some of their deep differences, the confrontation of an overall common danger is a very productive factor that is new in our situation. So I feel that the ecological crisis can actually move us forward.

ROCKEFELLER: Is it reasonable to think of the different religions as each and every one a pathway into a new common shared world of democratic and ecological social values? In other words, is the objective in the religions to bring people to a local place, or ultimately to a global place, which is the great community of life in which we all share?

THE DALAI LAMA: There's a close relation between various different religions. That I consider something very important. I feel that it is possible to develop such an understanding. According to my experience, as time goes by there have been many positive developments. There are two levels of relationship between the religions. One level is trying to see similarities. That's okay up to a certain point. Obviously there are also differences in philosophy and traditions. There are many differences, and some are fundamental differences. This involves another level of relationship, and I always am looking on that level. If different religions have a negative attitude toward one another, then it creates a lot of problems and also tragedy among humanity, doesn't it? From that point of view, one can see the enormous importance of having a close understanding between different religions.

Then there is another reality. First, on this planet, there are believers and nonbelievers. I think the nonbelievers are the majority. There are more than five billion human beings. I think the true religious believers are hardly one billion perhaps. Within that group, there are so many different philosophies or traditions of religion. Whether we like it or not, it's a fact that we have to coexist and live together. And at the same time, it is impossible to convert everyone to one religion. Also it is, I think, not advisable to make some kind of universal religion. Because then we'll lose the special and unique potential of each different religion. It is much better to preserve each individual religion.

Under those circumstances, if we take a wider perspective, there is every reason to be in harmony and to work together. Different philosophies and different practices are an individual business. So usually I call the different religions something like the supermarket of religion. That is far better than one single religion. Among humanity there are so many varieties of mental disposition. One religion cannot serve, or cannot satisfy the larger number of people. A variety of religions serves a much bigger portion of humanity.

So I feel that if we look realistically, and take a wider perspective, there is every reason we should come together, work together, respect each other, learn from each other. So I'm, in that respect, quite optimistic.

SCHORSCH: I think the real threat to religions working together comes from the rise of fundamentalism. I do believe that the ecological crisis is a positive factor that will compel many of us to sit around tables and pool our wisdom and resources. But the ecological crisis or the failure of technological society drives many to simple-minded solutions. That is a worldwide phenomenon. And so there is an embrace of the most appalling kind of fundamentalism. Now, that may be a repudiation of technology but it is also a highly uncooperative, divisive posture. I think it will make any kind of inter-religious cooperation practically impossible. So I believe that what we have to address as religious leaders is the widespread escape into simple-minded religious postures as a solution.

MCFAGUE: I want to support very much what Rabbi Schorsch has said. While we could see the ecological crisis as a common danger, unfortunately it often doesn't appear to be a danger. I mean, it is of greater proportions than a worldwide war, but we don't feel it that way. Look what the United States has done, deploying troops in a matter of weeks without a second thought in the Gulf situation. We are faced with something much more critical. But it's silent for the most part. Or we only feel it at certain times, like the summer of 1988. It's coming, of course, but we can put it to one side. Perhaps one of the most important things that the religious leaders can do, because the religions are in many ways the guardians of the values and worldviews of culture, is to make this common danger speak—be seen for what it is in its immensity. Those of us gathered here, although we see things in different ways, can at least agree that we are faced with an enormous danger, and that part of what we mean by salvation is the well-being of our planet and its creatures. Whatever else one wants to say about salvation, surely, one is talking about health, the health of our planet and all its creatures.

ROCKEFELLER: How does one go about making ethical decisions with reference to the environment, deciding what is good and bad, right and wrong? Reference has been made to historical revelation as being one basis for making these judgments. It's also true that there are many in our society today, particularly people working in the scientific realm, who adopt a pragmatic approach and look to the experimental method and causes and consequences to decide what actions are good or bad. Where does guidance come from in settling complicated environmental questions that have an ethical dimension to them?

ENGEL: Clearly one of the greatest dangers we face is a diminishment in cultural and local pluralism. It's as important that we preserve human, cultural, and religious pluralism as it is that we preserve bio-diversity, if evolution of life on this earth is to continue. Now, that suggests to me that we not only have to think globally and locally, but that we always have to be thinking finally in terms of a comprehensive principle of unity and variety. Whatever ethical choices and deliberations we engage in, they should be of many different kinds and should be fundamentally situational and experimental. That is to say, we need today a great variety of cultural and religious experiments in how to live well, self-sufficiently, sustainably in local ecosystems.

It seems to me that the UNESCO system of biosphere reserves, which is seeking an integration of cultural, social, political, economic, and ecological factors within discrete areas throughout the world; the bioregional movement as it has begun to have impact, particularly in advanced industrial countries; and the many experiments and movements within the rest of the world to reclaim the land where the community developed and where it has a future, are of the essence of the kind of ethical methodology that we need. In other words, the test is not only whether or not those local communities can sustain themselves, but whether they can participate sustainably in a reciprocal series of relationships with other communities throughout the world.

So my own criterion is to a large degree pragmatic. But the sources of those experiments, of course, are the deep resources and wells of wisdom within those communities, as well as within the modern sciences. I think it has to be extremely eclectic from that standpoint.

SCHORSCH: Since I really don't have a lot of wisdom, I'll reflect. I was struck this morning by His Holiness' presentation. No one at this table comes from a richer religious tradition than the Dalai Lama. And yet he spoke with all the fervor and substance of a humanist. Ultimately, when we come down to some of these questions it is our basic humanity that we draw on. The religious values may sensitize us, but ultimately there is a deep amount of pragmatism in our response to the crisis.

I think that one factor that we have to recognize in working toward the future is human fallibility. We are capable of creating such technological monsters that if something goes wrong, they will afflict an area of the globe with damage for eons. Now, it seems to me that things often go wrong because of us. Not because of technological flaws, but human flaws. And I wonder whether that reality ought not to stay our hand, ought not to induce us to make decisions not to build certain things. These supertankers. Or nuclear plants. One human error is enough to punish generations. Ought we to move into that kind of technological size, given our human limitations? That is a factor that I think society has not as yet been willing to take into consideration in making technological decisions. But I must say that it is a factor that gives me great pause.

ROCKEFELLER: I have another question, which has come from the audience: "Organized religions are often opposed to artificial birth control. Do you really think that it is realistic to expect to improve the human impact on the environment without artificial birth control?"

ENGEL: No, I absolutely don't think we're going to be able to improve the human lot without birth control.

SCHORSCH: I am deeply troubled by the predictions for depopulating the countryside. We're building ever larger megalopolises. Farmers are leaving the land. And we are not attempting to do anything in a policy way to reverse that. It seems to me that, if we're going to responsibly address the ecological crisis, we must begin to think about employing people differently and turning agriculture into a vast industry. Driving the small farmer off the land and building bigger cities is a formula for ecological disaster. I don't think that we as a nation have begun to articulate our fears on this issue.

ENGEL: It seems to me that the urban areas are precisely the centers for the battle. It's the Northern and Southern urban conglomerations which are placing the most tremendous drain upon the resources of this planet. If the cities can be changed, we will make a tremendous step forward in meeting the crisis. One cannot begin to think, when one is in the very heart of the city and in the most alienated, concrete skyscraper within it, that one is anywhere else except on this earth, and deeply connected to it. Our failure to understand that is, of course, the problem.

It seems to me that in addition to stopping the inflow of people from the hinterland into the cities, we must begin to take very seriously our urban centers as the sources, the actual centers for renewal of the planet, which means beginning to do agriculture in the cities. It means reforesting, as much as we can, the cities. It means decentralizing the cities. It means becoming aware of the connections and interdependencies of the cities and the rest of the world. It means creating the art in the cities that will in fact help us emotionally to live there as well as to remember the larger context in which we live. This is a very vital point.

ROCKEFELLER: We began by discussing the issue of personal reconstruction in one's own life, and a fair amount has been said about that. Let's look a little further now at the issue of social reconstruction and the formulation of a planetary agenda for social action. What are the one or two top priority items on the planetary agenda today?

NASR: This is not something that is simple to talk about. Since I'm myself a guest in this country, I'm most grateful that I'm able to live here under these difficult circumstances in my life during the last twelve years. But first of all, the highly industrialized countries must stop, if I can use a strong metaphor, sucking the blood of the nonindustrial part of the world. That is at the very, very top of the agenda. Everything else really is unfortunately secondary at the present moment.

ENGEL: I would suggest that in formulating such an agenda, we focus on those things that, if they happen, are irreversible. And in light of that criterion, I see a number of things that ought to be at the very top of the agenda. First, it seems to me that the population explosion is something that will be extremely difficult to reverse if it's allowed to continue. The prognostications of the United Nations are absolutely frightening. We are adding a hundred million people every year. This planet now has five billion, three hundred million human beings. In the year 2025, that's thirty-five years from now, we will have a population of eight and a half billion human beings. That is a situation which is irreversible and therefore has to be addressed immediately.

Second, the warming of the planet is a situation that, once it occurs, cannot be readily reversed. Therefore, the need to halt destruction of the rain forests, and replanting trees must come at the very top of our agenda.

Third, the problem with the ozone layer is also something that cannot be reversed once it's allowed to happen. Therefore, its destruction must be halted as quickly as possible. The level of response to this problem shows how even the United States can be moved along at an accelerated pace by the realization of the urgency of the crisis. And ultimately I think that's what will motivate nations and religions and individuals, the realization that this is a crisis of a magnitude we have never faced before.

SHENANDOAH: I would like to say again that it has to come from the individual. We have to cultivate that feeling for the earth in order to bring about these changes. It has to come from a feeling for the earth. Otherwise there will be no direction that is real. It will be someone telling someone else what to do. The power can be generated from the people, if it comes from the heart and from inside your being.

ROCKEFELLER: There are many more questions, I know, in your minds. My mind, too, is full of questions. We cannot bring this all to a neat conclusion. We've engaged an enormously complex problem in the course of the last three days. It is a problem, however, that we can all agree is a radical one. It requires major social changes. It means transforming our societies in ways that we are only just beginning to understand. In a document like *Caring for the World*, a planetary agenda has been proposed that is very helpful.

The environmental crisis is clearly a fundamental ethical problem because attitudes must be changed that have to do with the values by which we live. It is a religious problem because, in the final analysis, only a transformation of consciousness on the level of the heart, or what in the Asian traditions is called the heart-mind, is going to awaken the needed faith and motivation, and compassion and courage. There are great spiritual resources in the world religions and in other traditions that can be of benefit in this regard. It is also necessary for there to be a certain amount of adjustment and reconstruction within the various religions, if they are to respond fully and effectively to the environmental crisis.

We must continue to work on all of this collectively. The environmental crisis challenges us, in Ron Engel's words, to "keep faith with life" and to work cooperatively to develop the kind of local life-styles and world community that will ensure the well-being of all people and every life form on this planet.

EPILOGUE

Brooding over the Abyss

John C. Elder

Fig. 10. The two parts of Judith Anderson's (1935–) title for this etching, Missa Gaia: This is My Body *(1988) evoke the spiritual and emotional richness of her image.* Missa Gaia *relates the artwork to the ancient worship of the Goddess or Great Mother, antedating the patriarchal symbols of the divine which have been dominant in Western culture.* This is My Body *connects it with the Christian sacrament of Communion. Through the cycles of birth and death, and within the swirl of evolution, human beings participate in a web of life extending far beyond our individual existences. (Photo by Jim Colando)*

THE art exhibition that accompanied the "Spirit and Nature" symposium included a 1988 etching by Judith Anderson entitled *Missa Gaia: This is My Body*. In it, a female figure, her face cast downward, sits with her knees apart in birthing posture, surrounded by multitudes of creatures—pandas, eagles, butterflies, whales, grizzlies, mice, and elephants—that have emerged from her body. This image offers a contemporary interpretation of the Great Mother, who is also celebrated in the most ancient object in the exhibition, a shell figurine fashioned in Crete almost six thousand years ago. Today's renewal of interest in the Great Mother reflects a desire to balance the male symbols of the divine which have dominated Western history, and in doing so to heal the rift between nature and spirit that has often troubled our dealings with the earth and with each other.

Looking at Anderson's etching one also feels a powerful sense of brooding, as the woman, one hand suspended in front of her mouth, gazes down with lidded eyes into the heart of life. In her other hand she loosely holds the fluke of a sounding whale. It seems that she may soon let this animal go, in a dive toward extinction. Anderson has written about the effect of mingled celebration and grieving in her work: "Earth mass, mass for the earth, mass of the earth. It is the Great Mother's celebration: the earth, the grasses, the seas and the infinite variety of creatures *are* her body, incarnations of her Being and creativity. And all return home to her womb in death, dismemberment, extinction. In the undisturbed rhythm of earth, life and death are intertwined and balanced in a vast exchange of lives. . . .

The Great Mother of the print, surrounded by and filled with animals, embodies at once both celebration and profound grief and anger."[1]

Such troubled brooding in the midst of life's variety and beauty may be an especially appropriate image for our quest for a spiritual basis of environmental responsibility. As we begin the decade of the 1990s, we must acknowledge that, regardless of whatever corrections we now make in our destructive environmental practices, we are inevitably entering a time of mass extinctions, and of the destruction of wild habitats unprecedented in human history. Driven by our uncontrolled human population, such problems as global warming, lack of drinkable water, and famine will become impossible to ignore. Faced with wounds to the biosphere that will not be healed during our lifetimes or those of our children, we must undertake a process of creative grieving. Such grieving calls for recognition of all that has been lost through human carelessness. But sorrow and remorse can also lead us to a new maturity, fostering humility in our daily actions, as well as an enhanced sense of the preciousness of nature's many remaining gifts. Chastened, we may gain the gift of a diverse, though depleted, community, in whose recovery we may participate and with whose incarnations we may identify.

Another powerfully brooding piece in the "Spirit and Nature" exhibit was *Imagem de Minha Revolta* (*Image of My Indignation*) by the Brazilian sculptor Frans Krajcberg. Born in Poland, Krajcberg relates the destruction of Brazil's rain forests to the Nazi holocaust. He has fashioned the charred remains of a forest into an assembly with twelve stumps arrayed on the ground in front of a blackened, twisted log set on end. Rough pieces of ceramic embedded in the wood on both sides add to the looming presence, mysteriously suggestive of a crucifix, a human form, a totem pole.

Krajcberg's sculpture is testimonial art, calling the viewer's attention to an atrocity. This is a central part of the artist's vocation in a time of holocaust. As Krajcberg writes, "Art must be for participation. . . . Today, art for art's sake has no significance. My culture is nature. My work is nature. I must continue to defend our nature."[2] At the same time, if art, or religion, points only toward the horrifying, who can bear to contemplate such testimony for long? In its fullness, the process of grieving must also involve celebration. Elegiac hymns to vanishing beauty look beyond particular forms to the fountain from which our world flows and into which the spray of species and individuals subsides.

On the wall beside *Imagem* hung a gelatin silver print by the pioneering American photographer Harold Edgerton. Entitled *Atomic Bomb Explosion*, it records one instant in a 1952 nuclear test in the Southwest. The photograph captures a perfect globe of fire and vapor, poised between the moment of detonation and the upward rush into a mushroom cloud. Behind that blinding circle of pure force is the desert sky, black by contrast, while the tiny silhouettes of Joshua trees in the foreground place this holocaust in a human scale. It is an utterly terrifying image of the destructive power released, but not controlled, by modern science and technology. Yet a horrified onlooker also cannot help but respond to the beauty of this dramatic print. More than one person, looking at the dancing spangles of light within the conflagration's core, said that it made them think of the big bang and the origins of the universe. Others, because of a large, contrasting spot on the left side and the softly interfolded tones within the sphere, said that it evoked the fertilization and divisions of a single cell. At the moment of destruction, we can glimpse our origins and perceive energy swirling through a universe beyond our powers of destructiveness.

Can the monumental horrors of our day bring us back down to earth, opening our eyes to that web of tender relatedness in which the biosphere finds its integrity? This question challenges religions that have sometimes emphasized the grandeur of the divine at the expense both of the earth and of the majority of human beings. But Whitehead contrasts our Western tradition's massive, overwhelming images of God with "the brief Galilean vision of humility" that has "flickered throughout the ages, uncertainly." "It does not emphasize the ruling Caesar, or the ruthless moralist, or the unmoved mover. It dwells upon the tender elements in the world, which slowly and in quietness operate by love; and it finds purpose in the present immediacy of a kingdom not of this world. Love neither rules, nor is it unmoved; also it is a little oblivious as to morals. It does not look to the future; for it finds its own reward in the immediate present."[3]

Whitehead's sense of a tender voice in the Western tradition illuminates a hopeful possibility at the current moment of human evolution. The global environmental crisis does not call upon us to reject the legacy of civilization. Rather we are asked to attend to voices within our traditions which have not been sufficiently heard. This is the significance of such developments as renewed interest in St. Francis, or in that moment in the history of Buddhism when the influence of Taoism helped to transform a world-denying sense of emptiness into a celebration of nature's plenitude. We are reevaluating, realigning, even reconstructing our religious traditions, given the insights and needs of our day. This possibility for simultaneously grieving over our history's destructive errors and nurturing the birth of renewed traditions is captured in the double meaning of the word "brooding." It expresses both sorrowful contemplation and devotion to the development of fresh life. Milton captured both meanings in the invocation to *Paradise Lost*, where the Holy Spirit "Dovelike [sat] brooding o'er the vast Abyss / And mad'st it pregnant."

Nature writing, and the naturalist science with which it is connected, speak to the challenges of our evolutionary moment in ways that reinforce the wisdom of the artists and philosophers. With its roots in the Romantic revolt against mechanistic philosophies, American nature writing has emerged as a genre particularly important to the environmental movement. From Thoreau to Annie Dillard and Edward Abbey, we see a literature of reflective, lyrical essays, grounded in science and attuned to spiritual values. In the inclusiveness of their voices, such authors nudge us past our education's dualistic sense of science and the arts, the human and the natural. Thoreau asks in *Walden*, "Shall I not have intelligence with the earth? Am I not partly leaves and vegetable mould myself?"[4]

The life-sciences, too, as they move toward an emphasis on fieldwork and an interest in communities rather than in organisms in isolation, are recovering a wholeness of vision that recalls their origins in "natural history" and "natural philosophy." Though Darwin's theory of evolution ushered in a new age of specialized research, he himself represented a tradition of the classically educated, spiritually attuned naturalist. The concluding paragraph of his *Origin of Species* accords with the spirit of "Galilean tenderness" which Whitehead affirms within the religious traditions of the West:

> It is interesting to contemplate a tangled bank, clothed with many plants of many kinds, with birds singing on the bushes, with various insects flitting about, and with worms crawling through the damp earth, and to reflect that these elaborately constructed forms, so different from each other, and dependent upon each other in so complex a manner, have all been produced by laws acting around us. . . . There is grandeur in this view of life, with its several powers, having been originally breathed by the Creator into a few forms or into one; and that, whilst this planet has gone cycling on according to the fixed law of gravity, from so simple a beginning endless forms most beautiful and most wonderful have been, and are being evolved.[5]

Darwin's view, like Whitehead's and Thoreau's, is of a community, not a hierarchy. It is consistent with the Native American belief in the inextricability of human life from the rest of nature; with the tendency of modern physics to perceive dynamic patterns of relationship rather than discrete, solid objects; with the democratic models of government that have steadily supplanted more authoritarian regimes over the past two centuries of human history. Such a sense of community, though supported by much eloquent testimony within our tradition, also challenges much in our conception of land-ownership and in our exploitation of natural resources. For most of us in the West, integrating the values of natural harmony into our day-to-day life calls for a spiritual reorientation so fundamental as to be, at heart, a conversion.

The last image one saw in leaving the "Spirit and Nature" exhibition was the Apollo 17 photograph of *Planet Earth*. As familiar as this picture is, from its frequent reproduction on posters and book jackets, it still resists reduction to a cliché. It has been just eighteen years since we gained this vision of the earth as a whole, its surface swirling with luminous beauty against the velvet blackness of space. Such a picture "proves," in a visceral way, what we have long known but have still to realize adequately in our social choices: that our planet is round, beautiful, small (from a vantage point not so very far away), and fragile. It offers itself as a new icon for spiritual practice in an age of environmental crisis.

NOTES

1. Judith Anderson, Statement on "Missa Gaia: This is My Body," Fall Equinox, 1988.
2. *New York Times*, 17 October 1989.
3. Alfred North Whitehead, *Process and Reality* (New York: Free Press, 1969), 401.
4. Henry David Thoreau, *Walden and Civil Disobedience*, ed. Sherman Paul (Boston: Houghton Mifflin, 1960), 95.
5. Charles Darwin, *The Origin of the Species* (London: Penguin, 1985), 460.

APPENDIX

United Nations

World Charter for Nature

The General Assembly,

Reaffirming the fundamental purposes of the United Nations, in particular the maintenance of international peace and security, the development of friendly relations among nations and the achievement of international co-operation in solving international problems of an economic, social, cultural, technical, intellectual or humanitarian character,

Aware that:

a. Mankind is a part of nature and life depends on the uninterrupted functioning of natural systems which ensure the supply of energy and nutrients,

b. Civilization is rooted in nature, which has shaped human culture and influenced all artistic and scientific achievement, and living in harmony with nature gives man the best opportunities for the development of his creativity, and for rest and recreation,

Convinced that:

a. Every form of life is unique, warranting respect regardless of its worth to man, and, to accord other organisms such recognition, man must be guided by a moral code of action,

b. Man can alter nature and exhaust natural resources by his action or its consequences and, therefore, must fully recognize the urgency of maintaining the stability and quality of nature and of conserving natural resources,

Persuaded that:

a. Lasting benefits from nature depend upon the maintenance of essential

Copies of, and information on, the *World Charter for Nature* may be obtained from the United Nations Environment Programme, New York Office, Room DC2–0803, New York, NY 10017. Telephone: (212) 963-8138.

ecological processes and life support systems, and upon the diversity of life forms, which are jeopardized through excessive exploitation and habitat destruction by man,

b. The degradation of natural systems owing to excessive consumption and misuse of natural resources, as well as to failure to establish an appropriate economic order among peoples and among States, leads to the breakdown of the economic, social and political framework of civilization,

c. Competition for scarce resources creates conflicts, whereas the conservation of nature and natural resources contributes to justice and the maintenance of peace and cannot be achieved until mankind learns to live in peace and to forsake war and armaments,

Reaffirming that man must acquire the knowledge to maintain and enhance his ability to use natural resources in a manner which ensures the preservation of the species and ecosystems for the benefit of present and future generations,

Firmly convinced of the need for appropriate measures, at the national and international, individual and collective, and private and public levels, to protect nature and promote international co-operation in this field,

Adopts, to these ends, the present World Charter for Nature, which proclaims the following principles of conservation by which all human conduct affecting nature is to be guided and judged.

I. General Principles

1. Nature shall be respected and its essential processes shall not be impaired.

2. The genetic viability on the earth shall not be compromised; the population levels of all life forms, wild and domesticated, must be at least sufficient for their survival, and to this end necessary habitats shall be safeguarded.

3. All areas of the earth, both land and sea, shall be subject to these principles of conservation; special protection shall be given to unique areas, to representative samples of all the different types of ecosystems and to the habitats of rare or endangered species.

4. Ecosystems and organisms, as well as the land, marine and atmospheric resources that are utilized by man, shall be managed to achieve and maintain optimum sustainable productivity, but not in such a way as to endanger the integrity of those other ecosystems or species with which they co-exist.

5. Nature shall be secured against degradation caused by warfare or other hostile activities.

II. Functions

6. In the decision-making process it shall be recognized that man's needs can be met only by ensuring the proper functioning of natural systems and by respecting the principles set forth in the present Charter.

7. In the planning and implementation of social and economic development activities, due account shall be taken of the fact that the conservation of nature is an integral part of those activities.

8. In formulating long-term plans for economic development, population growth and the improvement of standards of living, due account shall be taken of the long-term capacity of natural systems to ensure the subsistence and settlement of the populations concerned, recognizing that this capacity may be enhanced through science and technology.

9. The allocation of areas of the earth to various uses shall be planned, and due account shall be taken of the physical constraints, the biological productivity and diversity and the natural beauty of the areas concerned.

10. Natural resources shall not be wasted, but used with a restraint appropriate to the principles set forth in the present Charter, in accordance with the following rules:

a. Living resources shall not be utilized in excess of their natural capacity for regeneration;

b. The productivity of soils shall be maintained or enhanced through measures which safeguard their long-term fertility and the process of organic decomposition, and prevent erosion and all other forms of degradation;

c. Resources, including water, which are not consumed as they are used shall be reused or recycled;

d. Non-renewable resources which are consumed as they are used shall be exploited with restraint, taking into account their abundance, the rational possibilities of converting them for consumption, and the compatibility of their exploitation with the functioning of natural systems.

11. Activities which might have an impact on nature shall be controlled, and the best available technologies that minimize significant risks to nature or other adverse effects shall be used; in particular:

a. Activities which are likely to cause irreversible damage to nature shall be avoided;

b. Activities which are likely to pose a significant risk to nature shall be preceded by an exhaustive examination; their proponents shall demonstrate that expected benefits outweigh potential damage to nature, and where potential adverse effects are not fully understood, the activities should not proceed;

c. Activities which may disturb nature shall be preceded by assessment of their consequences, and environmental impact studies of development projects shall be conducted sufficiently in advance, and if they are to be undertaken, such activities shall be planned and carried out so as to minimize potential adverse effects;

d. Agriculture, grazing, forestry and fisheries practices shall be adapted to the natural characteristics and constraints of given areas;

e. Areas degraded by human activities shall be rehabilitated for purposes in accord with their natural potential and compatible with the well-being of affected populations.

12. Discharge of pollutants into natural systems shall be avoided and:

a. Where this is not feasible, such pollutants shall be treated at the source, using the best practicable means available;

b. Special precautions shall be taken to prevent discharge of radioactive or toxic wastes.

13. Measures intended to prevent, control or limit natural disasters, infestations and diseases shall be specifically directed to the causes of these scourges and shall avoid adverse side-effects on nature.

III. Implementation

14. The principles set forth in the present Charter shall be reflected in the law and practice of each State, as well as at the international level.

15. Knowledge of nature shall be broadly disseminated by all possible means, particularly by ecological education as an integral part of general education.

16. All planning shall include, among its essential elements, the formulation of strategies for the conservation of nature, the establishment of inventories of ecosystems and assessments of the effects on nature of proposed policies and activities; all of these elements shall be disclosed to the public by appropriate means in time to permit effective consultation and participation.

17. Funds, programmes and administrative structures necessary to achieve the objective of the conservation of nature shall be provided.

18. Constant efforts shall be made to increase knowledge of nature by scientific research and to disseminate such knowledge unimpeded by restrictions of any kind.

19. The status of natural processes, ecosystems and species shall be closely monitored to enable early detection of degradation or threat, ensure timely intervention and facilitate the evaluation of conservation policies and methods.

20. Military activities damaging to nature shall be avoided.

21. States and, to the extent they are able, other public authorities, international organizations, individuals, groups and corporations shall:

a. Co-operate in the task of conserving nature through common activities and other relevant actions, including information exchange and consultations;

b. Establish standards for products and manufacturing processes that may have adverse effects on nature, as well as agreed methodologies for assessing these effects;

c. Implement the applicable international legal provisions for the conservation of nature and the protection of the environment;

d. Ensure that activities within their jurisdictions or control do not cause damage to the natural systems located within other States or in the areas beyond the limits of national jurisdiction;

e. Safeguard and conserve nature in areas beyond national jurisdiction.

22. Taking fully into account the sovereignty of States over their natural resources, each State shall give effect to the provisions of the present Charter through its competent organs and in co-operation with other States.

23. All persons, in accordance with their national legislation, shall have the opportunity to participate, individually or with others, in the formulation of decisions of direct concern to their environment, and shall have access to means of redress when their environment has suffered damage or degradation.

24. Each person has a duty to act in accordance with the provisions of the present Charter; acting individually, in association with others or through participation in the political process, each person shall strive to ensure that the objectives and requirements of the present Charter are met.

CONTRIBUTORS

John C. Elder

John C. Elder teaches English and Environmental Studies at Middlebury College, where he has developed a special interest in Natural History as a context for integrating the physical and social sciences with the humanities. His publications include *Imagining the Earth: Poetry and the Vision of Nature* (Urbana: University of Illinois Press, 1985), and *The Norton Book of Nature Writing* (New York: Norton, 1990), of which he was co-editor. He was one of the organizers of the Middlebury symposium on "Spirit and Nature: Religion, Ethics, and Environmental Crisis."

J. Ronald Engel

J. Ronald Engel is Professor of Social Ethics at Meadville/Lombard Theological School affiliated with the University of Chicago and an ordained minister in the Unitarian Church. He is the author of *Sacred Sands: The Struggle for Community in the Indiana Dunes* (Middletown, Conn.: Wesleyan University Press, 1983). He is also the co-editor with Joan Gibb Engel of *The Ethics of Environment and Development: Global Challenge, International Response* (London: Belhaven Press, 1990; rpt. Tucson, Ariz.: University of Arizona Press, 1990), which contains essays by twenty-one authors from fifteen countries. Professor Engel has served as the chair of the Ethics Working Group of the International Union for Conservation of Nature and Natural Resources (IUCN) in connection with the preparation of the second version of the World Conservation Strategy.

Tenzin Gyatso, His Holiness the 14th Dalai Lama

His Holiness the 14th Dalai Lama, Tenzin Gyatso, is the spiritual and temporal leader of the Tibetan people. Since 1960 he has resided in Dharmsala, the seat of the Tibetan Government-in-exile. He is the author of many books and essays, including *Kindness, Clarity, and Insight* (Ithaca, N.Y.: Snow Lion, 1984), *Ocean of Wisdom* (Santa Fe, N.M.: Clear Light Pub., 1989), and *A Human Approach to World Peace* (London: Wisdom, 1984). *The Dalai Lama, A Policy of Kindness:*

An Anthology of Writings By and About the Dalai Lama (Ithaca, N.Y.: Snow Lion, 1990) contains some of his most recent writings. He has lectured widely throughout the world and in 1989, he received the Nobel Prize for Peace and the Congressional Human Rights Award.

Sallie McFague

Sallie McFague is the E. Rhodes and Leona B. Carpenter Professor of Theology at the Divinity School of Vanderbilt University in Nashville, Tennessee. From 1975 to 1979 she served as the Dean of the Vanderbilt Divinity School, and she has taught at Smith College, the Yale Divinity School, and the Harvard Divinity School. A member of the United Methodist Church and a leading theologian in the English-speaking world, her many books and articles include *Literature and the Christian Life* (New Haven: Yale University Press, 1966), *Metaphorical Theology* (Philadelphia: Fortress, 1982), and *Models of God: Theology for an Ecological, Nuclear Age* (Philadelphia: Fortress, 1987). She approaches theology as an imaginative metaphorical process and draws upon the findings of astrophysics, cosmology, and biology in her efforts to develop an idea of God relevant to the late twentieth century.

Seyyed Hossein Nasr

Seyyed Hossein Nasr is currently University Professor of Islamic Studies at George Washington University in Washington, D.C. He has served as Professor of Philosophy and the History of Science at Tehran University, Chancellor of Aryamehr University, Iran, and the first president of the Iranian Academy of Philosophy. He has also taught at Harvard University, Temple University, and the American University of Beirut, and he is the first Muslim scholar to deliver the prestigious Gifford Lectures at the University of Edinburgh in Scotland. His publications include over twenty books, and among them are *Ideals and Realities of Islam* (New York: Harper Collins, 1989), *Islam and the Plight of Modern Man* (Kuala Lumpur: Foundation for Traditional Studies, 1982), *Man and Nature* (New York: Harper Collins, 1991), *Science and Civilization in Islam* (Cambridge, England: Islamic Text Society, 1987), *Sufi Essays*, 2d ed. (Albany: SUNY Press, 1991), *Three Muslim Sages* (Delmar, N.Y.: Caravan Books, 1976), and *Knowledge and the Sacred* (Albany: SUNY Press, 1989).

Robert Prescott-Allen

Robert Prescott-Allen is a writer and consultant on integrating environmental conservation and economic development. While Senior Policy Adviser for the International Union for Conservation of Nature and Natural Resources (IUCN), he organized and drafted the World Conservation Strategy. He is writer and senior consultant for *Caring for the Earth*, a second version of the World Conservation Strategy, which has been prepared by IUCN in collaboration with the United Nations Environment Programme (UNEP) and the World Wide Fund for Nature (WWF). Among his publications is a book co-authored with Christine Prescott-Allen: *The First Resource: Wild Species in the North American Economy* (New Haven: Yale University Press, 1986).

Steven C. Rockefeller

Steven C. Rockefeller is Professor of Religion and the former Dean of the College at Middlebury College. He is the author of *John Dewey: Religious Faith and Democratic Humanism* (New York: Columbia University Press, 1991), and co-editor of *The Christ and the Bodhisattva* (Albany: SUNY Press, 1987), a collection of essays that grew out of a 1984 Middlebury College symposium. His current research is focused on the integration in the contemporary world of democratic social values, ecology, and religious faith. He was the principal organizer and director of the Middlebury symposium on "Spirit and Nature: Religion, Ethics, and Environmental Crisis."

Ismar Schorsch

Ismar Schorsch is Chancellor and Professor of Jewish History at the Jewish Theological Seminary of America in New York City, the primary center in the Western hemisphere for the study of Judaism. As chancellor of the seminary, Dr. Schorsch leads a religious movement of 1.5 million Conservative Jews in the United States and 300,000 overseas. Dr. Schorsch has served as president of the Leo Baeck Institute, and his writings and translations focus on the historical development of Conservative Judaism, the history of European Jewry, and the emergence of historical consciousness in modern Judaism. He is the author of *Jewish Reactions to German Anti-Semitism, 1870–1914* (New York: Columbia University Press, 1972), and the translator and editor of *The Structure of Jewish History and Other Essays* (New York: Jewish Theological Seminary, 1975), by Heinrich Graetz. He has given public addresses on a variety of critical contemporary issues including environmental ethics.

Audrey Shenandoah

Audrey Shenandoah is the Cultural Supervisor of the Onondaga Indian School and an Elder of the Eel Clan of the Onondaga Nation, one of the six nations of the Haudenosaunee, also known as the Iroquois Confederacy. A respected teacher among her people, she has given public addresses on the traditional Native American understanding of the Sacred Cycle of Life, humanity's relationship to the earth, and environmental ethics. She has spoken before the United Nations in Geneva, and in January 1990 she was one of the keynote speakers at a conference on environmental issues in Moscow sponsored by the Global Forum of Spiritual and Parliamentary Leaders.

SELECTED BIBLIOGRAPHY

History, Philosophy, and Religion

Andrews, Valerie. *A Passion for This Earth*. San Francisco: HarperSanFrancisco, 1990.

Badiner, Allan Hunt, ed. *Dharma Gaia: A Harvest of Essays in Buddhism and Ecology*. Berkeley: Parallax Press, 1990.

Berry, Thomas. *The Dream of the Earth*. San Francisco: Sierra Club Books, 1988.

Buber, Martin. *I and Thou*. Trans. Walter Kaufmann. New York: Scribner's, 1970.

Buber, Martin. *Hasidism and Modern Man*. New York: Harper, 1958.

Callicott, J. Baird, and Roger T. Ames, eds. *Nature in Asian Traditions of Thought: Essays in Environmental Philosophy*. Albany: SUNY Press, 1989.

Capra, Fritjof. *The Tao of Physics*. Boston: Shambhala, 1975.

Cobb, John, Jr., and Charles Birch. *The Liberation of Life: From the Cell to the Community.* Cambridge, England: Cambridge University Press, 1981.

Dalai Lama. *Ocean of Wisdom*. Santa Fe, N.M.: Clear Light Publishers, 1989.

Dalai Lama. *Kindness, Clarity, and Insight*. Trans. and ed. Jeffrey Hopkins. Ithaca: Snow Lion, 1984.

Diamond, Irene, and Gloria Feman Orenstein, eds. *Reweaving the World: The Emergence of Ecofeminism*. San Francisco: Sierra Club Books, 1990.

Dooling, D. M., and Paul Jordan-Smith, eds. *I Become Part of It: Sacred Dimensions in Native American Life*. New York: Parabola Books, 1989.

Ehrenfeld, David, and Philip J. Bentley. "Judaism and the Practice of Stewardship." *Judaism* 34 (1985): 301–11.

Eliade, Mircea. *The Sacred and the Profane*. New York: Harper and Row, 1959.

Engel, J. Ronald. *Sacred Sands: The Struggle for Community in the Indiana Dunes*. Middletown, Conn.: Wesleyan University Press, 1983.

Fox, Matthew. *Breakthrough: Meister Eckhart's Creation Spirituality in New Translation*. New York: Image Books Doubleday, 1980. Reprint. New York: Doubleday, 1991.

Fox, Matthew. *The Coming of the Cosmic Christ*. San Francisco: Harper and Row, 1988.

Fox, Matthew. *Original Blessing*. Santa Fe: Bear & Company, 1983.

Fromm, Erich. *To Have or To Be?* New York: Harper and Row, 1976.

Gadon, Elinor W. *The Once and Future Goddess*. San Francisco: Harper and Row, 1989.

Hargrove, Eugene C., ed. *Religion and Environmental Crisis*. Athens: University of Georgia Press, 1986.

Harris, Monford. "Ecology: A Covenantal Approach." *CCAR Journal* 23 (1976): 101–8.

Hart, John. *The Spirit of the Earth: A Theology of the Land*. New York: Paulist Press, 1984.

Hughes, J. Donald. *American Indian Ecology*. El Paso: Texas Western Press, 1983.

Joranson, Philip N., and Ken Butigan, eds. *Cry of the Environment: Rebuilding the Christian Creation Tradition*. Santa Fe: Bear & Company, 1984.

Kaufman, Gordon D. *Theology for a Nuclear Age*. Philadelphia: Westminster Press, 1985.

Linzey, Andrew, and Tom Regan, eds. *Animals and Christianity: A Book of Readings*. New York: Crossroad Publishing Co., 1988.

McFague, Sallie. *Models of God: Theology for an Ecological, Nuclear Age*. Philadelphia: Fortress Press, 1987.

Macquarrie, John. *In Search of Deity: An Essay in Dialectical Theism*. London: SCM Press, Ltd., 1984.

Merchant, Carolyn. *The Death of Nature: Women, Ecology and the Scientific Revolution*. New York: Harper and Row, 1989.

Merchant, Carolyn. *Ecological Revolutions: Nature, Gender, and Science in New England*. Chapel Hill: The University of North Carolina Press, 1989.

Moltmann, Juergen. *God in Creation: A New Theology of Creation*. New York: Harper and Row, 1985.

Nash, Roderick Frazier. *Wilderness and the American Mind*. Rev. ed.: New Haven: Yale University Press, 1973.

Nash, Roderick Frazier. *American Environmentalism: Readings in Conservation History*. New York: McGraw Hill, 1990.

Nasr, Seyyed Hossein. *Islam and the Plight of Modern Man*. 1975; 2d ed., Kuala Lumpur: Foundation for Traditional Studies, 1982.

Nasr, Seyyed Hossein. *Man and Nature: The Spiritual Crisis of Modern Man*. Kuala Lumpur: Foundation for Traditional Studies, 1968. New ed. New York: Harper Collins, 1991.

Neihardt, John G. *Black Elk Speaks: Being the Life Story of a Holy Man of the Oglala Sioux*. Lincoln: University of Nebraska Press, 1988.

Nhat Hanh, Thich. *Present Moment, Wonderful Moment*. Berkeley: Parallax Press, 1990.

Nhat Hanh, Thich. *Peace Is Every Step: The Path of Mindfulness in Everyday Life*. Ed. Arnold Kotler. New York: Bantam Books, 1991.

Roberts, Elizabeth, and Elias Amidon. *Earth Prayers from Around the World*. San Francisco: HarperSanFrancisco, 1991.

Robinson, J. A. T. *Exploration into God*. Stanford: Stanford University Press, 1967.

Ruether, Rosemary Radford. *Sexism and God-Talk: Toward a Feminist Theology*. Boston: Beacon Press, 1983.

Santmire, H. Paul. *The Travail of Nature: The Ambiguous Ecological Promise of Christian Theology*. Philadelphia: Fortress Press, 1985.

Schweitzer, Albert. *Out of My Life and Thought*. New York: Holt, 1949.

Shapiro, David S. "God, Man and Creator." *Tradition* 15 (1975): 25–47.

White, Lynn, Jr. "The Historical Roots of Our Ecologic Crisis." *Science* 155 (10 March 1967): 1203–7. Reprinted in Ian Barbour, ed. *Western Man and Environmental Ethics*. Reading, Mass.: Addison-Wesley, 1973.

Zukav, Gary. *The Dancing Wu Li Masters: An Overview of the New Physics*. New York: William Morrow & Co., Inc., 1979.

Environmental Ethics

Attfield, Robin. *The Ethics of Environmental Concern*. New York: Columbia University Press, 1983.

Berry, Wendell. *The Unsettling of America*. San Francisco: Sierra Club Books, 1977.

Callicott, J. Baird. *In Defense of the Land Ethic: Essays in Environmental Philosophy*. Albany: SUNY Press, 1989.

Dalai Lama. *A Policy of Kindness: An Anthology of Writings by and about the Dalai Lama*. Ed. Sidney Piburn. Ithaca, N.Y.: Snow Lion, 1990.

Daly, Herman E., and John B. Cobb, Jr. *For The Common Good: Redirecting the Economy toward Community, the Environment, and a Sustainable Future*. Boston: Beacon Press, 1989.

Engel, J. Ronald, and Joan Gibb Engel, eds. *Ethics of Environment and Development: Global Challenge and International Response*. London: Belhaven Press, 1990. Published by arrangement with Belhaven Press, Tucson: University of Arizona Press, 1990.

Hargrove, Eugene C. *Foundations of Environmental Ethics.* New York: Prentice-Hall, 1988.

Helfand, Jonathan I. "The Earth is the Lord's: Judaism and Environmental Ethics." In *Religion and Environmental Crisis*, 38–52. Ed. Eugene C. Hargrove. Athens: University of Georgia Press, 1986.

Leopold, Aldo. *A Sand County Almanac*. Oxford: Oxford University Press, 1949.

Midgley, Mary. *Animals and Why They Matter*. Athens: University of Georgia Press, 1984.

Momaday, N. Scott. "An American Land Ethic." In John G. Mitchell, ed., *Ecotactics: The Sierra Club Handbook for Environmental Activists*. New York: Simon and Schuster, 1970.

Nash, Roderick Frazier. *The Rights of Nature: A History of Environmental Ethics*. Madison: University of Wisconsin Press, 1989.

Regan, Tom. *The Case for Animal Rights*. Berkeley: University of California Press, 1983.

Regan, Tom, and Peter Singer, eds. *Animal Rights and Human Obligations*. Englewood Cliffs, N.J.: Prentice-Hall, 1976.

Rolston, Holmes. *Environmental Ethics: Duties to and Values in the Natural World*. Philadelphia: Temple University Press, 1988.

Rolston, Holmes. *Philosophy Gone Wild: Essays in Environmental Ethics*. Buffalo, N. Y.: Prometheus Press, 1986.

Schumacher, E. F. *Small is Beautiful: Economics as if People Mattered*. New York: Harper, 1973.

Singer, Peter. *Animal Liberation: A New Ethics for Our Treatment of Animals*. New York: Avon Books, 1975.

Environmental Activism

Andruss, Van, *et al.*, eds. *Home! A Bioregional Reader*. Philadelphia: New Society Publishers, 1990.

Brown, Lester R., *et al.*, eds. *State of the World 1991: A Worldwatch Institute Report on Progress Toward a Sustainable Society*. New York: Norton, 1991.

Christensen, Karen. *Home Ecology: Simple and Practical Ways to Green Your Home*. Golden, Colo.: Fulcrum Publishing, 1990.

Devall, Bill. *Simple in Means, Rich in Ends: Practicing Deep Ecology*. Salt Lake City, Utah: Peregrine Smith Books, 1988.

Devall, Bill, and George Sessions. *Deep Ecology: Living as if Nature Mattered*. Salt Lake City, Utah: G. M. Smith, 1985.

The EarthWorks Group, eds. *50 Simple Things You Can Do to Save the Earth*. Berkeley: EarthWorks Press, 1989.

The EarthWorks Group, eds. *The Next Step: 50 More Things You Can Do to Save the Earth*. Kansas City, Mo.: Andrews and McMeel Books, 1991.

The Global Tomorrow Coalition. Ed. Walter H. Corson. *The Global Ecology Handbook: What You Can Do About the Environmental Crisis*. Practical Supplement to the PBS Series "Race to Save the Planet." Boston: Beacon Press, 1990.

Hynes, H. Patricia. *Earth Right: Every Citizen's Guide*. Rocklin, Calif.: Prima Publishing, 1990.

Little, Charles E. *Greenways for America*. Baltimore: The Johns Hopkins University Press, 1990.

McKibben, Bill. *The End of Nature*. New York: Random House, 1989.

Meeker-Lowry, Susan. *Economics as if the World Really Mattered*. Santa Cruz, Calif.: New Society Publishers, 1988.

Naar, Jon. *Design for a Liveable Planet. How You Can Help Clean Up the Environment*. New York: Harper and Row, 1990.

Newkirk, Ingrid. *Save the Animals! 101 Easy Things You Can Do*. New York: Warner Books, 1990.

Piasecki, Bruce, and Peter Asmus. *In Search of Environmental Excellence: Moving Beyond Blame*. New York: Simon & Schuster, 1990.

Report of the President's Commission: *Americans Outdoors: The Legacy, The Challenge*. Washington, D.C.: Island Press, 1987.

Rifkin, Jeremy, ed. *The Green Lifestyle Handbook: 1001 Ways You Can Heal the Earth*. New York: Holt, 1990.

Schell, Jonathan. *The Fate of the Earth*. New York: Avon Books, 1982.

Steger, Will, and Jon Bowermaster. *Saving the Earth: A Citizen's Guide to Environmental Action*. New York: Knopf, 1990.

World Commission on Environment and Development. *Our Common Future*. New York: Oxford University Press, 1987.

World Conservation Union (IUCN), United Nations Environment Programme (UNEP), and World Wide Fund for Nature (WWF). *Caring for the Earth: A Strategy for Sustainable Living*. Gland, Switzerland, 1991. Available from Island Press, Covelo, CA 95428.

World Resources Institute, in collaboration with United Nations Environment Programme and United Nations Development Programme. *World Resources, 1990–1991*. New York: Oxford University Press, 1990.

Nature Writing

Anderson, Lorraine, ed. *Sisters of the Earth*. New York: Random House, 1991.

Bly, Robert, ed. *News of the Universe: Poems of Twofold Consciousness*. San Francisco: Sierra Club Books, 1980.

Dillard, Annie. *Pilgrim at Tinker Creek*. New York: Bantam Books, 1974.

Ehrlich, Gretel. *The Solace of Open Spaces*. New York: Viking, 1985.

Emerson, Ralph Waldo, and Henry David Thoreau. *Nature/Walking*. Ed. John C. Elder. Boston: Beacon, 1991.

Finch, Robert. *The Primal Place*. New York: Norton, 1983.

Finch, Robert, and John C. Elder, eds. *The Norton Book of Nature Writing*. New York: Norton, 1990.

Giono, Jean. *The Man Who Planted Trees*. Chelsea, Vt.: Chelsea Green Publ. Co., 1985.

Hay, John. *In Defense of Nature*. Boston: Little, Brown, 1969.

Hoagland, Edward. *Walking the Dead Diamond River*. San Francisco: North Point, 1985.

Lopez, Barry. *Arctic Dreams*. New York: Scribners, 1986.

Matthiessen, Peter. *The Snow Leopard*. New York: Viking, 1978.

Mills, Stephanie, ed. *In Praise of Nature*. Covelo, Calif.: Island Press, 1990.

Snyder, Gary. *The Practice of the Wild*. San Francisco: North Point, 1990.

Snyder, Gary. *Turtle Island*. New York: New Directions, 1974.

Thomas, Lewis. *The Lives of a Cell*. New York: Bantam, 1974.

Van Matre, Steve, and Bill Weiler. *The Earth Speaks*. Warrenville, Ill.: The Institute for Earth Education, 1983.

INDEX